The Private Lives of Animals

A Chanticleer Press Edition
Design and Coordination by
Massimo Vignelli and
Gudrun Buettner

The Private Lives

McGraw-Hill Book Company
New York St. Louis
San Francisco Auckland
Bogotá Johannesburg
London Mexico Montreal
New Delhi Panama
São Paulo Singapore Sydney
Toronto

Commentary by
Roger Caras

# of Animals

First published by Grosset & Dunlap, Inc., New York, 1974
Second printing by McGraw-Hill Book Company,
1221 Avenue of the Americas, New York, N.Y. 10020

ISBN 0-07-009794-1

Library of Congress
Cataloging-in-Publication Data
Caras, Roger A.
The private lives of animals.
Reprint. Originally published: New York: Grosset & Dunlap,
1974.
1. Animals—Behavior—Pictorial works. I. Title.
QL751.C35   1987   591.51′022′2   86-27669
ISBN 0-07-009794-1

Printed and bound in Hong Kong

Prepared and produced by
Chanticleer Press, Inc., New York

Cover photograph: A pair of lions mating. [Guenter Ziesler]
Overleaf: Salmon in the Adams River, British
Columbia: the beginning of life. [Steven C. Wilson]

If we correctly interpret cave paintings and early fetish figures, man's interest in animals lies at the very roots of human culture. There apparently has never been a time when animals have not held endless fascination for man and when he was not trying to express what he felt on the subject. Animals are at the root of our graphic, literary and performing arts, they are in our religion, and they are the very stuff of our mythology. They are base rock in our folklore and an essential element in children's literature. Animals are everywhere in human awareness. The average adult or child can tell you more about the wildlife of Africa than he can about the ethnology or geography of the continent. He may not know the capital of Alaska but he will be able to hold forth on grizzly bears and salmon runs.

It is perhaps fruitless for us to assay the relationship between man and animal. All we know is that the interest is nearly universal and ever increasing. Witness these facts: more people will go to the zoo this year than to all spectator sporting events combined. Although there are other attractions in our national parks besides wildlife, it is significant that the number of people entering the parks exceeded 182,000,000 last year. And one out of every 363 Americans visited the game parks of either Kenya or South Africa in that same year.

People seek out animals in zoos, museums and in wild areas simply because they are inherently fascinating. They represent a profound mystery, the ability of a species to find the best way to live in a given set of circumstances no matter how harsh or bizarre it may seem to us. Within that master mystery there are lesser ones. There is evolution, the eternal quest for a better solution to the problems of survival and the building of that solution into a species until the species itself emerges as a new and different animal. There is behavior—actions that evolve just as physical features do. There

Introduction

is birth, a miracle repeated billions of times every day, but still a profound and bewildering miracle. And no less wonderful is the behavior of some parent animals once birth has taken place. As we move through the following pages we must remember that all of the animals in the photographs are doing the same thing: seeking as long a life as they can gain for themselves and, unwittingly, for their species. An animal's existence is a delicate balance between risk and safety, danger and countering action, death and life. We must never lose sight of that fact as we travel the world in company with our greatest wildlife photographers in each section of this book. The problem changes—cryptic behavior, food-getting, mating, maternal care—but in each instance we will be seeing animals doing what is best for them, given the circumstances and the life style evolved thus far by their species.

Our knowledge of animal life is many, many times what it was even a few generations ago. The freshman biology student today knows more about the science of animal life than scientists did little more than a hundred years ago. He may not have the field experience, but the books, the magazines, the audio-visual aids and the concepts available to him would have been unbelievable to his great-great-grandfather. Yet, we still know relatively little about animals. We do not understand instinct—not really—and we are beginners when it comes to knowing inheritable characteristics. Ethology—the evolution of behavior—is a new science and it is only in recent years that we were equipped to comprehend the relationship one species has to all others, plant as well as animal. People such as Konrad Lorenz and Niko Tinbergen have helped reveal new horizons, and a brilliant generation of young field observers—such as George Schaller and Jane Van Lawick-Goodall, for example—are helping us find the dimensions of those horizons. Hundreds of men and women around the world are afield with ever more sophisticated photographic equipment, recording animal behavior in terms that are both artistically and scientifically valid. This is a good time for the enquiring mind.

Ironically, even as we assimilate the accumulated knowledge of hundreds of observers over the past centuries, the very creatures we seek to study are disappearing all around us. Over a thousand animal forms (speaking of vertebrates alone) are in some degree of danger. The rate at which animals are vanishing accelerates every year. Our fascination with the mystery of animal life is a race against the clock. It is also a test on some kind of ultimate moral plane, for if we allow ourselves to lose that race we shall have been swamped by one more animal mystery—the behavior of man.

Despite the popular conception of reproduction, it may or may not be bound up with sex. In myriad lower invertebrates, eggs and sperm are not involved in reproduction, or are involved in some generations and not others. Sea squirts can reproduce sexually or by simply budding, and the adult animals that result from the two processes are indistinguishable.

Marine worms are another case of alternate means. They may produce eggs and sperm at one time and yet reproduce at another time by simply breaking up, each part then developing into a full-fledged adult. Jellyfish, too, reproduce by means of fertilized eggs or by growing miniature versions of themselves *on* themselves. Of course, living things produced by asexual reproduction have a uniform genetic inheritance, except from those new forms that arise through gene mutation.

Among the vertebrate animals, however, before an individual can join the mainstream of life, generally two others of its kind must meet, usually engage in a ritualistic acceptance of each other, and then contribute their genetic substance. In some cases, notably among fish, the eggs and sperm are simply broadcast and there is no actual meeting of individuals.

There are, thus, endless variations on the theme of reproduction and even a simple listing of them would be impossible here. In many of the higher vertebrates pair-bonding precedes the actual mating. Pair-bonding involves mate selection, courtship rituals, the act of copulation, and it can involve nest building and site selection. In many species it is a life-time arrangement; in others it is temporary.

There are many ways in which animals form pair-bonds. There are monogamous species in which a male and a female join and permit no intrusion upon their union for as long as it lasts. There are polygamous species where the male mates with more than one partner (many males, including seals, some antelopes, and horses, assemble harems), and there are polyandrous species where one female accepts more than one male; finally, there are species that are both polygamous and polyandrous. Whatever the pattern, it seems certain that in a given species the style that has evolved is the one that best facilitates reproduction and therefore offers the most promise for the future of the species.

In species where pair-bonds are formed, courtship and display are of great importance. They are usually forms of visual and auditory communication very specific to each kind of animal. They help synchronize reproductive rhythms, making sure that both animals reach the necessary plateaus at the same time, and ultimately facilitating fertilization.

Once two animals select each other and a bond begins to form, two essential social processes become evident. First, there is a concentration of response patterns: animals that once responded to many others of their species now respond only to each other, and any natural aggression toward each other becomes more restrained. At the same time, either member of the pair, or both, may become more aggressive toward others of their species that venture too close at this critical time.

Hybrids, the mixture of two species, are relatively rare in nature. A species that regularly hybridized would be in danger of vanishing and being replaced by a new species, and it would certainly reduce the efficiency of the reproductive process by confusing genetic inheritance. The more complex the pre-mating behavior, the more elaborate the song or posturing, the less likely it is that different species will mate. Seemingly minute differences in the colors of two otherwise similar species may also provide identification factors that help prevent species from mixing.

Male birds sing to establish their mating territory and to attract a partner. Peacocks fan their tails and bighorned sheep

Courting and Mating

butt their heavy-horned heads together with a crash that can be heard for miles. Bull wapiti bugle their challenge to other males until the valleys ring with their power and proclamations. In a pond a frog sings and puffs himself up, displaying his splendor to another of his kind. At 10,000 feet a male eagle whirls through an incredible aerial display, working himself into a pre-mating frenzy while his aerobatics are paralleled far below by a fragile winged insect who may hatch in the morning, mate at midday and die before the sun sets.

On other levels, animals reproduce in ways that can only be described as totally impersonal. The amoeba dividing in two, the self-fertilizing worm, the fish that discharges its milt over the eggs of females all seem to be acting automatically, without any sense of individuality. It is simply the species going forward. It is not at all as romantic and exciting as we like to project, not always as conscious an act, but it is a constant theme—the duplication of self in kind and the resultant forward motion of the species.

Overleaf, preceding page: His tail feathers in full display, a peacock (Pavo cristatus), *a distant relative of our domestic chicken, parades before a prospective mate. His incredible tail and the ease with which these birds can be domesticated have made the common peafowl one of the most familiar exotic birds in the world. Our language is replete with anthropomorphic misconstructions such as "proud as a peacock." If anything, he is pleading with the drab female to pay attention to him. If she does approve of his movements and display, she will become his mate. [Eric Hosking] His throat sac inflated, an American toad* (Bufo americanus), *above, announces the start of its breeding season. These animals cannot actually copulate since the male lacks a penis. In a process known as amplexus the male will crawl up on the female's back and fertilize her eggs as she deposits them. [Karl H. Maslowski]*

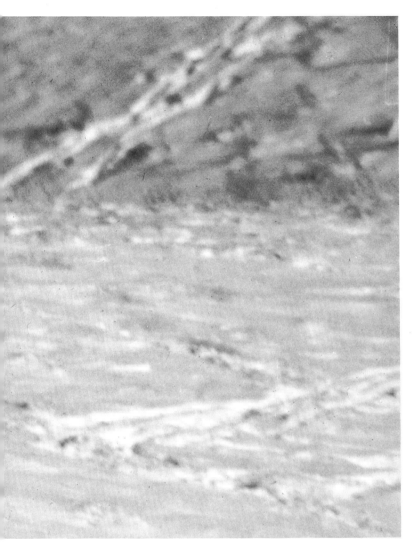

Smell can be of immense importance in mating in creatures as varied as butterflies and sheep. The young bighorn ram at right (Ovis canadensis) sniffs the air anxiously as he smells an ewe in heat. [Leonard Lee Rue IV] Above, he pursues her vigorously across a snowy slope. Having attracted him with her trail, the ewe works him into a frenzy with her ritualized avoidance behavior. Their autumn mating will assure the birth of a lamb in spring, when food is plentiful and the weather reasonably mild. [Leonard Lee Rue III]

For some animals mating is a mass display—a swarming nuptial flight precisely timed to season and hour. The midges (order Diptera), *left*, are such creatures. Beating their wings more than a thousand times a second, they appear by the myriads for their mating flight. It is soon over, but another generation has been assured.

Spiders have a complex reproductive system. The male has no penis and must deposit semen in the female's genital orifice with his feelers. The male garden spider (Argiope), *above*, approaches a nearly blind female seeking her undivided attention. Popular writers often refer to the way female spiders kill their mates after or during mating; it does happen but not nearly as often as some writers have suggested. [*Both by David Cavagnaro*]

Western grebes (Aechmophorus occidentalis), *right*, engage in courtship display. It is performed, ballet-like, by both sexes. At its completion a pair-bond will be established and procreation

can get under way. [*Jen & Des Bartlett: Bruce Coleman, Inc.*]
Cattle egrets (Ardeola ibis), *far right, engage in a somewhat subdued*
display. Male and female perch on a branch, raising and lowering
their feathers, sometimes for hours at a time. Each species has a
unique pattern of pre-mating display, thus discouraging
hybridization. [*Thase Daniel*]

17

The Adélie penguin (Pygoscelis adeliae), *left, is a very gregarious animal. A million or more may assemble in a rookery during the breeding season. The pebbles with which they construct a crude nest are scarce in this land of perpetual snow and ice. The male finds one, often by stealing it from another's nest, and offers it to the female. If she accepts it, the pair is bound, and they stand belly-to-belly crooning their mating song. [Francisco Erize] Among the pinnipedia, or fin-footed animals, there is total dominance by the males. The gray seals* (Halichoerus grypus), *above, exhibit a marked difference between the sexes (called sexual dimorphism) that supports this dominance—a sharp contrast in size. But in many fish and most spiders the female is the larger animal. [Fred Bruemmer]*

The average South American bull sea lion (Otaria byronia), *above, has about ten cows in his harem. As the Antarctic summer approaches, the herds break up into breeding and nonbreeding groups. Any time after her fourth year the cow may be claimed by a bull who will be much larger than she is. Playful nips and other sexually stimulating acts are all part of the ritual.*
[*Francisco Erize*]

The ocelots (Felis pardalis), *like most cats, engage in vigorous pre-mating play. Small bites, usually around the face and head, stimulate both cats. Sometimes tempers flare, but that, too, is part of it. There appears to be a very fine dividing line between sexual stimulation and pain. Rarely, though, is any harm done.* [*Thase Daniel*]

Sound as well as visual display is important in many species. The wandering albatrosses (Diomedia exulans) *hiss, wheeze and rattle their bills, upper left, the male shrieks his enthusiasm, lower left, and the display reaches its climax, above, as both birds, bills to the sky, "yell," with their wings widespread. This frantic exercise tells each one that the other is ready. [Niall Rankin]*

Overleaf : A pair of black rhinoceros (Diceros
bicornis) *mating. These ton to a ton-and-a-half animals mate at any
time of the year. Males often fight when a cow nearby is in heat,
and a female may attack a bull before mating with him. These
rhinos are rough animals, usually solitary except for mothers with
their young calves. The males establish a territory of sorts,
marking it with dung heaps, urine and rubbing sites.*
[B. Leidmann: Bavaria-Verlag]

The avocet (Recurvirostra americana), *a large shorebird, engages in an elaborate mating ritual. At left, the female assumes the mating position, with her chin and bill almost touching the water, her whole attitude one of anticipation. The male preens and walks in front of her, moving closer and closer until he finally brushes her tail. He may throw water on her head. He then mounts her, above. At right, the sexual act completed, the male shrouds his mate with a wing as they cross bills. They will take several steps forward and then move away from each other. Their parting is temporary, for in this species the male helps incubate the eggs.* [Ed Bry]

American coppers (Lucaena helloides), *above, copulating after a courtship display by the male butterfly that included impressive aerobatics.* [*Larry West*]

The remarkable photograph at right corroborates a description long ago by the French entomologist Jean Henri Fabre of how the female mantis (Mantodea) *may actually consume her mate during the sexual act. Although his head and thorax are already eaten away by the female, the male shown here continues to copulate. He will stop only when she reaches his abdomen.* [*Peter Ward: Paul Popper Ltd.*]

Nudibranchs, left, among the most colorful of all invertebrates, are hermaphroditic marine molluscs with both male and female organs present in each individual. Self-fertilization is possible, but the genetics of the species is better served by the act seen here— two Claucus *engaging in reciprocal copulation, the male organ of each joining with the female organ of the other. [William M. Stephens: Tom Stack & Associates]*

Despite the rather cumbersome shell with which they must maneuver, amber snails (Succinea putris), above, manage to perform the sexual act. Their union lasts longer than the corresponding act among most vertebrates. [J. Markham: Bruce Coleman, Inc.]

The tropical lizards known as geckos are gregarious and generally nocturnal. During the mating season a male will grab the first lizard it meets. If it happens to be another male, a brief tussle will ensue; if it is a female, she will submit quietly. Here a male

(Palmatogecko rangei), top, *grasps a female as they mate. Fearsome puff adders (Bitis arietans),* top, *mate in a rather conventional manner. In nonvenomous species the males sometimes bite the females while mating. Among venomous species this is avoided: even though there may be a degree of immunity to the venom of their own species, the long fangs could be very damaging: venom needed for use against prey would be lost, and a reaction to a large dose of venom might occur. [Both by Edward S. Ross]*

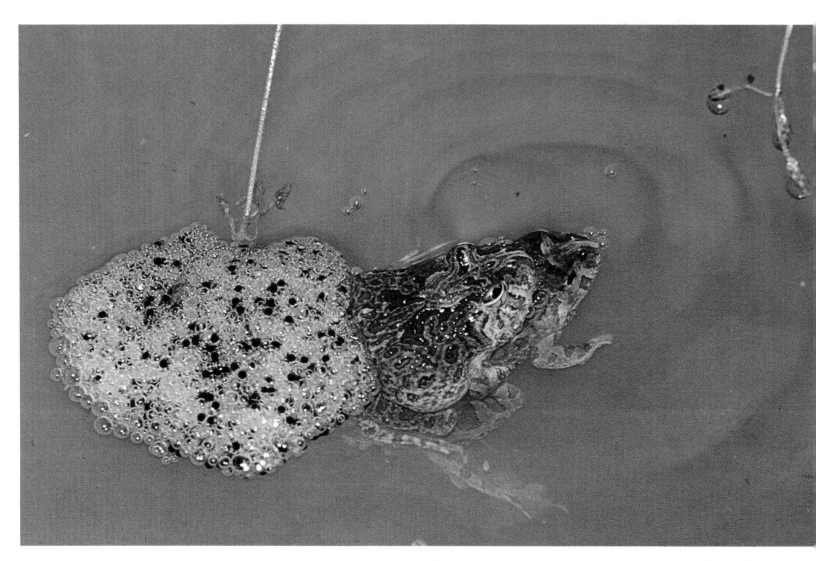

Despite appearances, the ornate burrowing frog (Limnodynastes ornatus), *above, and the Irangi night frog* (Afrixalus *spp.*), *right, are not copulating since frogs are not equipped for such an act. Instead, the female is releasing eggs and the male is releasing sperm; the fertilized eggs will develop outside the female's body. Were their breeding activities not coordinated with the spring, cold-blooded animals like frogs could not survive. Thus they are ready to breed as soon as they come out of hibernation.* [Stanley Breeden; Edward S. Ross]

Higher animals, including all mammals, must copulate in order to reproduce. In some species, such as the great red kangaroo, left, the Masai giraffe, center, and the African elephant, above, the act appears to the human observer to be a triumph of instinct over anatomy. [Left and center, Tierbilder Okapia; right, Norman Myers: Alpha Photo Associates]

Male lions are polygamous and breed throughout the year. In fact, in a well-established pride, where females do the killing, breeding is about all the male lion does. The male lion at left will not bite down on the female as he seems about to do. He is rough, but his mate has a more violent temper, and he will not go too far. Above, having completed their union, the lions rest. They may mate again after a nap or the male may attend to the next female in the pride's hierarchy. That will depend, though, on the female. If she wishes to copulate again it will be dangerous for a lesser female to approach the male; a female higher on the pecking order could displace her. [*I. W. Sedgwick: Animals Animals*]

Life is such an all-encompassing experience that the fact that there are points in time where it actually starts and stops is bewildering. Of the two, of course, the starting is the more beautiful experience. That is the beginning of a miracle. Animals are born in various ways and at different levels of development. An amoeba divides and becomes two and, in a rather broad sense of the word, we can say two amoebas have been born. Certainly, the original amoeba has gone. Many fish and reptiles, all amphibia, all insects, all birds and some few unusual mammals produce eggs and are called oviparous. The other higher animals produce living young; they are viviparous.

Newborn or newly hatched animals display a wide range of development even among closely allied species. Rabbits are helpless, blind and hairless at birth while the related hares are precocious, that is, highly developed. Among rodents, mice are pink and helpless whereas guinea pigs are up and about almost immediately, eyes open, and fully furred.

All this is true of birds as well and we have a fine vocabulary to cover the matter. Some newly hatched chicks are covered with down (ptilopaedic), active and quite able to hold their own (precocial), and leave the nest almost immediately (nidifugous). Others are naked (psilopedic), helpless (altricial), and confined to the nest for some time (nidicolous).

The method of birth, the time, the place and the condition of the young are all determined by external pressures. In the very long view they may be seen as responses, the ways that a species can best duplicate itself and survive. In some cases weather is the determining factor and so, very often, is food supply. Migratory creatures—salmon, for example, and many birds— must rely on reproductive cycles that are precisely timed. An animal that faced a flight of five or six thousand miles between summering and wintering grounds could not survive as a species if individuals received the inner signals to mate at the wrong end of the trip. If the salmon laid its eggs at the seaward end of its journey instead of in the sweet water of its own natal stream, the salmon would cease to be.

Environmental and ecological pressures also dictate the number of young that are born to individuals of each species. A female sea turtle may produce a hundred eggs but few of her young will survive their first hours; gulls, fish, raccoons, feral dogs, hungry human beings and a variety of other predators and scavengers steal the eggs or gobble up the hatchlings before or just after they reach the water's edge. However, one hundred eggs ensure that enough offspring survive to perpetuate the species. (What would happen, though, if every rat, lion or elephant gave birth to a hundred young at a time?) Elephants take extremely good care of their young (as opposed to turtles who never see theirs) and the infant mortality is low. One baby elephant, or occasionally two, at a time are enough. In some areas cub mortality among lions may reach 50 percent because of parasites and, occasionally, cannibalism. But the lion population of an area must be held down so as not to put too much pressure on prey animals; so a litter of lions ranges from two to four.

All of these patterns are the result of responses to ecological pressures over long periods of time. For the animals that survive it is the beginning of all the things we will view in this volume. Not all of the elements are there at birth, but the genetic potential is. A kangaroo, little more than an embryo at birth, emerges without its hind legs developed; they will develop as the joey feeds and nestles in the warmth of its mother's moist pouch.

Similarly, a baby elephant will one day uproot great trees and an Arctic tern will be able to migrate over 20,000 miles a year. Virtually all that the animal can ever be must be packed into it at birth.

Nests, Eggs and Birth

Changes in a species come about not as a result of environmental pressures during the life span of any single animal but through genetic mutations. Those mutations which survive—because they change the form and behavior of the animal, rendering it better able to utilize the opportunities within its environment—become part of the species. Strangely, under some circumstances this evolution can occur very rapidly; in bacteria, for example, or even in some insects, it can be accomplished in one generation. Evolution is not necessarily a long-drawn-out affair covering eons. It occurs in spurts; and sometimes on a grand scale.

Overleaf, preceding page: A northern gannet (Sula bassana) *at Bonaventure, Gaspé Peninsula, carries material to build a nest. These large sea birds always nest in dense colonies with immediate access to the sea. Intricate social patterns allow them to pack their nests into small areas without doing harm to the chicks.* [*Leonard Lee Rue IV: National Audubon Society*]
A fifteen-spine stickleback male (Spinachia spinachia) *passing through his nest while secreting threads that bind the nest together. When the nest is complete he will drive a gravid female into it and she will lay her eggs there.* [*Douglas P. Wilson*]

*Some animals mate and give birth in solitude, but not the flamingos.
Two subspecies are shown here: above are American flamingos*
(Phoenicopterus ruber ruber) *on Inagua in the Bahamas, and the
female turning her egg at right is the European flamingo*
(Phoenicopterus ruber roseus). *Some birds require a large
territory when they mate and nest, but not these highly gregarious
species. Only in a solid mass can they feel secure and go through
the reproductive sequence.* [Left, Sandy Sprunt: National Audubon
Society; right, Kenneth Fink: National Audubon Society]

Birds utilize every conceivable kind of nesting site. The kittiwakes
(Rissa tridactyla), *left, and the delicate fairy tern* (Anous albus)
or white noddy, above, have solved their common problem in
quite different ways. The former have chosen a cliff overlooking the
sea and utilized every ledge while the latter has chosen an
astonishingly precarious perch—the completely unprotected
junction of two branches in a small tree on Midway Island.
[Eric Hosking; David Cavagnaro]
Overleaf: The exquisite gelatinous egg mass of a Hexabranchus
imperialis *undulates in the sea. The Hexabranchus is a*
nudibranch, a marine mollusc without an external shell.
These very ancient animals have been depositing their cradles of
future nudibranchs in the sea for hundreds of millions of years.
The living chemistry of each egg is virtually identical to that of
every other. Change in the sea is slow, because it is a relatively stable
environment. [René Catala, Aquarium de Nouméa]

A variety of birth styles: upper left, an Australian insect with parental instincts. Although many people refer to all insects (and arachnids) as "bugs," the true bugs belong only to the insect order Hemiptera. *Here a hemipteran,* Canto parentum, *guards its eggs. When they hatch, the same attention will be given the young. Many insects simply lay a multitude of eggs and abandon them.* [Michael Morcombe]

A female black-widow spider (Latrodectus mactans), *lower left, one of the most venomous spiders, hovers near her egg mass. A predator approaching the eggs would face lethal opposition.* [Jack Dermid]

The pipefish (Dunckerocampus catala), *upper center, like its relative the sea horse, relegates the care of the young to the male. This apparent reversal of roles is not uncommon in the animal kingdom.* [René Catala, Aquarium de Nouméa]

A male water bug (Belostoma), *lower center, with eggs. The female*

placed the eggs on the male's back and then abandoned them to his charge. [*Alexander B. Klots*]

Two parasitic wasps (Megarhyssa), *above, use their slender ovipositors, or egg tubes, to deposit their eggs in the burrows of* Tremex, *horntail sawflies. When they hatch, the wasp larvae will feed on the* Tremex *larvae.* [*Edward R. Degginger*]

These Ridley turtles (Lepidochelys) *migrated across vast stretches of ocean to come ashore on the beach where they themselves were hatched. They are only a fraction of the number that hatched that day. Most never reached adult size before being taken by predators. A female, at right, deposits her many leathery eggs in a hollow she has scooped out of the sand with her hind flippers. After laying as many as one hundred eggs she will cover them again and leave it to sun and time to do the hatching.* [*David Hughes*]

Reptiles and fishes can be either oviparous (egg-layers) or viviparous (live-bearing) but almost all amphibia such as the pair of Agalychnis *frogs, above, are oviparous. At right, developing* Agalychnis *eggs. [Nicholas Smythe]*

Although this seagull chick (Larus), *above, long ago left its mother's body as a fertilized egg, this is the moment of birth. The chick has now completed the easiest phase of its existence: up to this point it has not been required to contribute to its own survival. Although at first helpless, as the days pass it will depend more and more for survival on its own patterned responses.* [*Andrew Hoglund: FPG*]

An American crocodile (Crocodylus acutus), *at right and on following page, emerges from its egg. As an adult it will be a powerful predator, but at the moment it is vulnerable to many other predators ranging from adult crocodiles to large fish and even birds. Some mammals may also prey on these small saurians in their first year.* [*All by Caulion Singletary*]

Above, Nile crocodiles (C. niloticus) *hatch. These eggs have survived the large monitor lizard whose population in many parts of Africa is now out of balance and whose depredations are a major threat to crocodile survival.* [*Jen & Des Bartlett*]

After a gestation period of over 350 days, the birth of a Burchell's zebra (Equus burchelli) *is a time of extreme danger for mare and foal alike. In their habitat of savannah grasses live the lion, the hyena, packs of Cape hunting dogs and the leopard. If a zebra foal doesn't stand within minutes, feed almost at once to gain energy, and isn't ready to move away with its mother in case of danger, it will quickly become a meal for a predator. In the event of attack, neither the herd nor the mother will be able to defend it.* [*Norman Myers*]

We don't really know what an instinct is. The term is a catch-all, a convenient basket into which to put some obvious as well as some questionable eggs.

This much is known: built into some animals is a whole complex of drives, urges and deep-rooted insistencies called the "family instinct." We may not fully understand how it works or how it is transmitted, but in observing wildlife it is obvious when it is present.

Most animals do not have the urge to band together or care for their young. Even in groups where such an urge generally occurs it may be missing in some species. Most birds, for example, are very attentive to their chicks; yet the megapods, or brush-turkeys, of Australia never so much as glance at their chicks once they are hatched in the huge mounds or incubators built by the adults. Some insects lay eggs and either die or move away without looking back. Other insects do protect their eggs and their young. The social insects—termites, wasps, bees and ants—tend to form a community brood in an extraordinary system of mutual benefit.

Most reptiles lay their eggs and abandon them to sun, rotting vegetation and fate. Not so the world's largest and one of its most dangerous venomous snakes. The magnificent king cobra, which may grow to be almost nineteen feet long, builds a nest by dragging dirt and vegetable matter into a heap with its considerable coils. Into the middle of this mass the female lays her eggs and then coils around the whole to guard them.

Salmon lay their eggs and die (the Pacific salmon) or return to the sea (the Atlantic salmon), yet other fish such as the mouth-breeders closely guard their young and even allow them to retreat into their cavernous mouths in time of danger.

A lioness or tigress will fight ferociously in defense of her cubs, while a guinea pig barely gives its young any attention.

In many species, it's the male's responsibility to care for the young, or at least to spend as much time at it as the female does. Male seahorses hatch the eggs, and male birds of many species work hard on the nest, share in the incubation of the eggs and make as many trips a day as the female to stuff the gaping mouths of the chicks.

Generally, the higher an animal is on the evolutionary scale, the less developed it will be at birth. Animals that retain juvenile characteristics for prolonged periods are exposed to their parents, or at least one parent, for those same prolonged periods. Generally that allows more time for teaching and learning and greater capacity for performance in a wider variety of situations. We must, however, put limits on that statement. It is clearly true of wolves, lions and elephants but clearly untrue of wasps, ants and bees. Like much else in the study of animals, truisms must be broached with caution.

Not all so-called family relationships involve parents and young. Wolves use "baby-sitters," unattached adults who watch over the young while their parents hunt. An old cow well beyond the breeding age may function within an elephant herd as an "auntie," and be as solicitous of a calf as its own mother. Older siblings also help care for young elephants. A father lion can be very patient with his cubs (he can also get testy and kill them) while a female grizzly bear will drive her mate away with terrible fury if he even approaches his own cubs (because he is likely to be a cannibal).

The only constant, though, seems to be this: when the miraculous and mysterious collection of drives we identify as "family instinct" exists, it is all-powerful. Such a bonding pervades all activities, dictates movement and presses members (some members) of a family or of a group together with a cohesiveness that ensures that species a greater chance for survival.

The Family Instinct

*Overleaf, preceding page: A cry from a young alligator will
sometimes arouse adults nearby into action whether or not they are
related to the youngster. Thus, despite appearances here, the big
American alligator* (Alligator mississipiensis) *may not be a parent
of the young one that has crawled onto its back but only a convenient
platform that just happens to be an adult of the same species.*
[*Les Line*]
*Some arachnids, like some insects, deposit their eggs and
abandon them. Others, like this wolf spider* (Lycosa *sp.*), *care for
the newly-hatched young to the point of transporting
them wherever they go.* [*Andreas Feininger*]

In contrast to the egg-laying alligator, the lion (Panthera leo), *above, and the spotted hyena* (Crocuta crocuta), *at right, carry their young around inside of them for approximately three and a half months. Both give birth to absolutely helpless infants who would perish in hours if not tended to constantly. These females display a powerful parental drive that continues until the young are independent.* [Norman Myers: Bavaria-Verlag; Norman Myers: Bruce Coleman, Inc.]

Variations on the theme of motherhood among mammals: At far left, a female gibbon (Hylobates spp.), classified among the great apes, carries her young with her as she swings through the trees. Her baby instinctively clings to her fur. [Edmund Appel] A beaver (Castor canadensis), upper left, a rodent, nurses a few of her two-week-old offspring while she grooms another. Grooming is vital to an animal whose fur must be insulation against cold water. [Jen & Des Bartlett: Bruce Coleman, Inc.] At lower left is a tamandua, or collared anteater (Tamandua tetradactyla), carrying its young. [Warren Garst: Tom Stack & Associates] Related to the tamandua, is the three-toed sloth (Bradypus griseus), above, an edentate, which is known for the way it moves about as it hangs from branches, exhibiting the same impulse to protect its infant young as shown by the far more athletic and apparently more responsive gibbon to the left. [Carl W. Rettenmeyer] At right is the vervet monkey (Cercopithecus pygerythrus), an exceedingly alert and active species, with its young. Babies are seldom released to move around and then only when the mother is certain no danger lurks nearby. [Edward S. Ross]

An immature kangaroo, called a joey, retreats to its mother's pouch. Member of the marsupials, a group found mainly in New Guinea and Australia, this great red kangaroo (Macropus rufus) *is among the most primitive of mammals. Some marsupials have a pouch into which the young find their way after a gestation of only forty days. As soon as her young is born, the female breeds again, but the development of the embryo stops after the 100-cell stage. It begins developing again only when its mother can properly care for it.* [*Alex Milich: Black Star*]

Although they live in the sea and have essentially torpedo-shaped bodies, the Pinnipedia ("fin-footed ones") are mammals. Here three females demonstrate their mammalian nature by suckling their young, supplying them with very rich milk. At upper left is a harbor seal (Phoca vitulina) *and her youngster, and at lower left, at the edge of the ice and ready to dive for safety, is a harp seal* (Pagophilus groenlandicus) *and her nursing pup. [Both by Fred Bruemmer] Far from the polar regions is a Hawaiian monk seal* (Monachus schauinslandi), *above, nursing her hungry offspring. She belongs to one of the very few tropical species of Pinnipedia. [David Cavagnaro]*

At left, a Virginia white-tailed deer (Odocoileus virginianus) *cleans a fawn just six minutes old. The baby, very vulnerable to predators, will not long be allowed the luxury of resting. It must recover quickly from the shock of birth and be prepared to flee. Above, the fawn is seven minutes old and has begun to nurse. At right, the new animal is already showing signs of interest in its surroundings. Part of the doe's task during the fawn's first weeks of life is to teach it the parameters of fear, for fear is the deer's greatest defense.* [Leonard Lee Rue III]

Overleaf: Emperor penguins (Aptenodytes
forsteri), *nearly four feet tall, lay their eggs and brood their young
during the Antarctic's severest season. There is no nest and the
single egg is held on top of the parent's feet by a pouchlike fold of
abdominal skin. Later the chick is brooded there, and until it is
quite large the parent shuffles for short distances, carrying its
chick along. During the incubation and brooding period the adults
may fast for as long as three months. Still later the half-grown
chicks will be gathered into nurseries, or creches, and be led down
to the sea.* [*Michael C. T. Smith: National Audubon Society*]

At left, on a stream bank in England, a mute swan (Cygnus olor) *rests with her young. Swans are extremely protective and both the female and male (cob) will attack anything that even approaches the cygnets. Attacks on full-grown men are common under these circumstances. Relatively few animals are equipped to stand up to the determined assault of a swan. [Jane Miller] On an island off Antarctica, a gentoo penguin chick* (Pygoscelis papua), *above, seeks cover under its mother, but will no longer fit. The mother tolerates the frantic shoving. Chicks of many species are endlessly demanding. [Niall Rankin] At right, in a heron rookery in the Camargue, France, a little egret* (Egretta gazzetta) *hen feeds her chicks. At this stage, the chicks' demands for food simply cannot be satisfied and the adults work to the point of exhaustion to still the chicks' complaint. [H. W. Silvester: Rapho Guillumette]*

Above, a Rainbow bee-eater (Merops ornatus) *catches an insect on the wing. The insect will be carried to the nest and fed to the peeping chicks. The process will be repeated again and again from dawn to dusk. The hours worked, the miles flown, the energy expended are prodigious. There is no gratitude—just another generation.* [*Michael Morcombe*] *At right, a male lovely wren* (Malurus amabilis) *has just fed one chick and now the other is squawking for a share. The female, in the meantime, is off hunting for more food.* [*Stanley Breeden*]

Whatever our human projections may be, sex is not the most compelling force in the life of animals; food-getting is.

When mating time comes, it may override all other considerations for the moment, but inevitably the individual animal must return to its primary requirement—food. It might be said that sex is to the species what food is to the individual—a means of surviving.

Appetites vary tremendously. Snakes may feed at weekly intervals whereas some shrews will starve to death in seven hours.

The feeding habits and requirements of an animal reflect its metabolism, its size and the efficiency of its digestive system. The elephant has to eat around the clock, consuming as much as four hundred pounds of vegetable matter a day because of its highly inefficient digestive system.

Some animals have extremely versatile diets. A hyena, as both hunter and scavenger, will take almost any protein food. Yet an animal like the koala bear (a marsupial and not a bear) must rely exclusively on the old leaves of a few species of gum tree. The young leaves of the same tree would kill it and leaves of related species would be useless to it as food.

We tend to pass judgement on animals because of their eating habits. A vulture is "disgusting" because it eats carrion, yet we eat cheeses that have become very "high." The wolf or bobcat that takes a newborn fawn is called "cruel," yet we eat veal. The hunting dogs that rip a fleeing hoofed animal apart are a "vicious pack of killers," yet we are just as carnivorous, eating live clams and oysters and boiling lobsters alive. All of these evaluations are obviously nonsense. If we were truly moral about eating we should all become vegetarians overnight. Unfortunately, a creature with the digestive system of a lion cannot; they cannot digest raw vegetable fibres in bulk.

The basic food of the animal kingdom is the vegetable kingdom. Animals eat plants. Some, like the herbivores, do it directly.

Other animals, like the lion and the tiger, do it indirectly by allowing a converter to act as a middleman. We must never forget that if all the plants on earth died tomorrow, all the animals would follow in short order.

Some animals, like man, are omnivores. Bears (except for the polar bear, which is nearly totally carnivorous) eat anything and so do some wild canids. Coyotes and certain foxes are extremely fond of melons and wild grapes. We seem to understand and forgive the omnivores more readily than we do the carnivores. There is an impression that some animals are meat-eaters; but they are really animal-eaters. There is a tremendous difference between the two. A lion or a wolf would suffer from malnutrition if fed exclusively on sirloin steaks. Animals that eat animals do so literally—hair, hide, meat and gristle, bone, undigested milk in the stomachs of young, material in the gut of an adult—all of this supplying traces of minerals we can only guess at. This applies as well to the carnivores that man has chosen to be his closest companions—dogs and cats. When we treat them as meat-eaters instead of animal-eaters we cheat them and shorten their lives.

Almost from the start some young animals can eat what their parents eat. Others cannot, and mammals have solved that problem with the mammary glands that have given them their name. Birds do not produce their own baby food, so they often create it out of the food they themselves eat. They swallow food and either retain it or partially digest it before regurgitating it for their young. Wolves and wild dogs, interestingly enough, also do something similar. A wolf on a kill may swallow large chunks of meat and by a means not clear to us it can then retard digestion and later cast up food for its cubs or mate.

Some animals feed each other. The female lion kills and the male lion helps himself; the Cape hunting dogs actually share

The Quest for Food

their kills. And food-offering is commonplace as part of the pre-mating ritual.

Some animals, like the browsers and grazers, usually have sufficient vegetation to feed on and they spend most of their lives doing just that. Many species have to travel to follow the seasons—winter and summer or dry and wet—but they do not work hard for their living. Other animals, like the cheetah and the leopard, must depend on immense skill and huge bursts of energy to get enough food to stay alive. All animals, when viewed from the evolutionary vantage point, are products of their dietary needs: the giraffe's neck, the lion's teeth, the hyena's bone-cracking jaws, the cow's stomach, the sucking instinct in young mammals, the owl's talons and the eagle's beak—all are diet-oriented. An animal is not only what it eats, it is designed so that it can eat.

Overleaf, preceding page: On Isle Royale a pack of timber wolves (Canis lupis) *circles a moose* (Alces alces). *If they sense that the cost in life and injury to themselves will be too great, they may linger for a few minutes and then leave. When the remains of a wolf kill are examined, it is very often found that the prey was sick or elderly. A moose, the largest member of the deer family, can weigh as much as 1800 pounds, so even in the fourteen-to-one confrontation seen in this aerial photograph, the moose has at least a weight advantage over the whole pack. Although there is no clear correlation between the weight of predators and prey, it can be a significant factor.* [L. David Mech]
Above, Cape hunting dogs (Lycaon pictus) *have culled a whitebearded wildebeest* (Connochaetus taurinus) *from its herd and are in the process of pulling it down. Unlike cats, hunting dogs are not equipped to make a clean, deft kill; they must tear the animal down.* [Norman Myers: Bruce Coleman, Inc.]

Cape hunting dogs run a small herd of Burchell's zebra (Equus burchelli). *They will select one animal out of the herd and cooperate to bring it to earth. A lone hunting dog could not begin to tackle the species' normal prey, but as a member of a magnificent coordinated hunting pack it can dine on all but the largest plains animals. [Norman Myers: Bruce Coleman, Inc.]*

A leopardess (Panthera pardus), *above, has made her kill and retreated to a high platform with it. She will store the carcass in the fork of the tree and eat at her leisure. Lions are unlikely to follow a leopard into a tree (although they can climb trees and do so in at least two locales in East Africa) and hyenas can't. On the ground the leopard might have to fight to keep its prize.*

A lion, at right, doesn't have to retreat with its prey. Any lion still agile and powerful enough to kill is probably capable of defending it against all comers except stronger lions. Lions normally eat near where the kill is made partly because they often take zebra and wildebeest, and dragging such heavy animals would entail a great expenditure of energy. A tired cat might lose its prey to one that had not been recently taxed. [Both by Norman Myers: Bruce Coleman, Inc.]

Pelicans eat some crustaceans but their principal diet is fish. The pelican at left has spotted some fish near the surface and drops down for the catch. The impact of its heavy body striking the water is thought to stun the fish long enough to make them easy prey. Above a bird has just penetrated the surface and seized a fish. It will toss the fish back into the pouch and hold it there until it can surface. [*Both by Kurt Severin*] On the right, *another brown pelican* (Pelicanus occidentalis) *breaks the surface with a fish which it will work up toward its throat and swallow whole. In Africa, pelicans hunt cooperatively, working in a semicircle and driving small fish into shallow water so that they can be scooped up.* [*Jen & Des Bartlett*]

An Atlantic puffin (Fratercula arctica), *above, with its catch. These birds nest by the thousands on cliffs overlooking the sea. They must feed often because they can cope only with small prey. Thus, they come and go constantly on their fish-hunting forays during daylight hours. Once they clamp down on a fish it has little chance of escaping.* [Les Line]

Baby turtles fall prey to seemingly unlikely predators. At right, a newly hatched Ridley (Lepidochelys) *is taken by a ghost crab.* [*David Hughes*]

We are apt to look on snakes as more uniform in appearance than most groups of animals familiar to us. There is a greater apparent difference between such birds as an ostrich and a hummingbird or between a rhinoceros and a shrew than there is between even a python and a cobra. Still, snakes are diverse and have developed life styles as different as their coldbloodedness will allow.

Above, an African egg-eating snake (Dasypeltis scabra) *swallows an egg. It will take the snake about five minutes to get the egg into a position where bony processes on the underside of its spine can puncture the shell and allow the contents to be swallowed. The empty shell will be cast up.* [*Kurt Severin.*]

At right, a corn snake (Elaphe guttata) *constricts a white-footed mouse* (Peromyscus leucopus) *preparatory to swallowing it. Contrary to popular belief, the snake will not crush its prey. Each time the mouse exhales, the snake will tighten its coils a little more. The mouse will soon become unconscious for want of oxygen. The snake may wait until the heart stops beating before commencing to swallow its prey head first. It is sensitive enough to feel its prey's heart and know when it has stopped.* [*Jack Dermid*]

A rhinoceros viper (Bitis nasicornis) *swallows a rat that it has injected with a venom powerful enough to incapacitate or even kill a man. Since the snake cannot chew, the venom contains an enzyme that aids in the digestion of its prey. Because they can be killed by rat bites and because even minor injuries may prove traumatic, such snakes wait for their venom to take effect.* [Edward S. Ross]

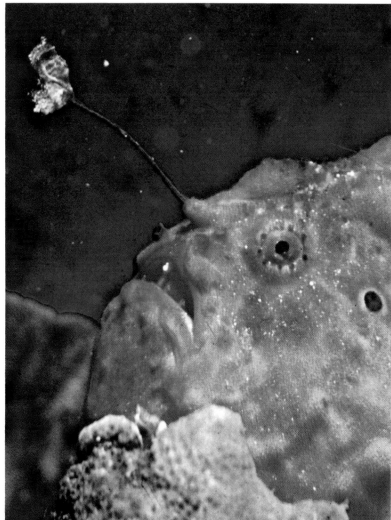

Two special adaptations for food-getting: the chameleon,
Chameleo dilepis, *above, used its tongue to retrieve a grasshopper.
The tongue can be flicked out more than the length of the
chameleon itself (tail included) in about 1/25 of a second. The
club-shaped end of the tongue is made adhesive by a secretion from
mucous glands. [Alan Blank: National Audubon Society]
An Angler fish (Antennarius commersoni), *at right, offers a
facial process as a lure. These special adaptations obviate the need
for active hunting and enable slow and awkward swimmers to
remain well fed. [René Catala, Aquarium de Nouméa]

Insect predators at work: at left, a praying mantis (Mantidae) eating a moth. Mantids are active hunters attacking insects of many kinds and other small animals of insect size. Their activities are essentially beneficial to man. [*Anthony Bannister: Natural History Photographic Agency*]

A blue mud-dauber wasp (Chalybion californicum), *above, uses its stinger to paralyze a spider. Wasps use their stingers without harm to themselves whereas many bees, like the honey bee, use their stinger only in defense and usually die because the stinger and its attendant organs and glands rip free from the bee's body.* [*Edward R. Degginger*]

At right, in South Africa, the larva of the giant ant lion (Palpares inclemens), *which has been lurking beneath the surface of the sand, reaches up with its efficient jaws and seizes a grasshopper.* [*Anthony Bannister: Natural History Photographic Agency*]

Various animals utilize flower products for food and in turn carry pollen that helps the plants reproduce themselves. At left, a white-lined sphinx moth (Celerio lineata), feeds on a penstemon flower. Its mouth is a long sucking-tube that enables it to get nectar and pollinate tubular-shaped flowers. [David Cavagnaro] Above is a honey possum (Tarsipes spenserae) from southwestern Australia. These little animals use their long snout and a tongue that is bristled at the tip to obtain nectar and pollen from flowers. They also eat some insects. At right, a yellow-faced honeyeater (Meliphaga chrysops) drinks nectar from a Christmas bell flower. Hummingbirds are the best known of the birds that have become adapted to this special diet. [Both by Michael Morcombe]

103

Caterpillars, left, devastating the leaf of a ti-tree. Caterpillars of
various species can be tremendously destructive to food and decorative
plants. Butterflies and moths, much admired by man, are at
one stage of life among the most costly to man. [Jen & Des Bartlett:
Bruce Coleman, Inc.]
A water vole (Arvicola terrestris), center, eating vegetable matter.
Such herbivorous animals are converters providing
plant-derived benefits to the carnivores that prey on them. [Stephen
Dalton: Natural History Photographic Agency]
A Galápagos Islands tortoise (Geochelone elephantopus), above,
is another animal that must depend on the plant kingdom for all
of its food needs. It eats prodigious quantities daily but in
undisturbed areas lives in balance with its habitat. [William E.
Ferguson]

A black or prehensile-lipped rhinoceros (Diceros bicornis), *above, browsing on a tree. These fascinating relics of the Pleistocene may weigh as much as two tons yet they are peaceful browsing animals. They can sometimes be short-tempered when surprised, but their reputation for being ferociously bad-tempered is groundless.* [Bruce Coleman, Inc.]

At right, another giant vegetarian—the African elephant (Loxodonta africana). *A vast amount of vegetable matter will be consumed each day by this bull, but no more than half of it will be thoroughly digested. It is fortunate that few animals require the amount of food needed by this bull. In a natural situation, with migratory patterns uninterrupted, elephants and other animals live in a working relationship to the habitat.* [Alpha Photo Associates]

5

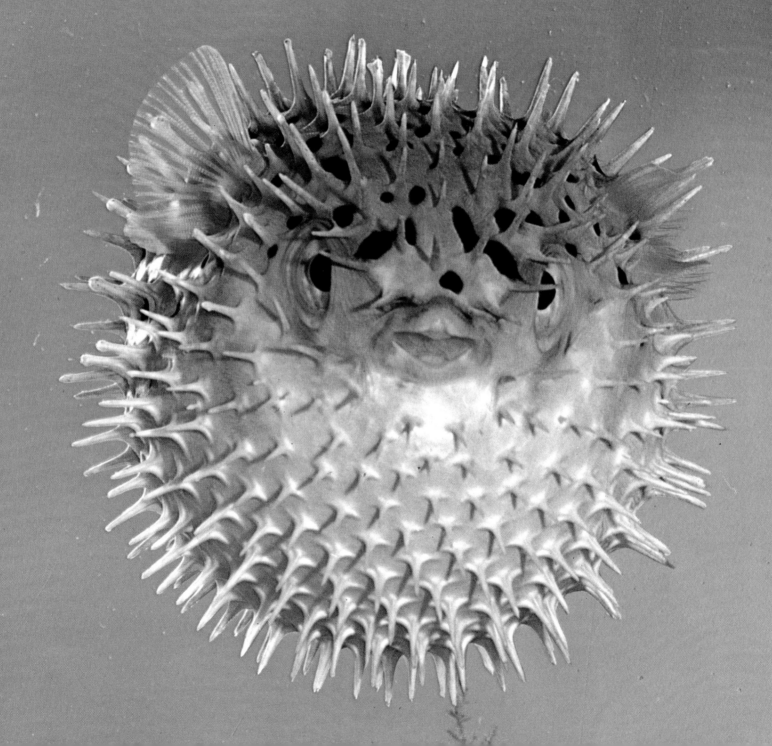

Not long ago I accompanied a small group down the steep western escarpment of Tanzania's most famous caldera—Ngorongoro Crater. As we made our way across the crater floor one man remarked how well fed and sleek the zebra and the wildebeest looked. "They are all in magnificent shape," he observed.

I replied, "Do you know what they call a zebra who is *not* in magnificent shape? Supper."

A few minutes later we came upon a scene that underscored the point. Forty spotted hyena were chasing each other around the edge of the great crater lake, each trying for its share of two wildebeest calves that had been run down and torn apart.

A lone hyena was pounding through the shallow water at the lake's edge trying to single out a flamingo from among the thousands that waded there.

The point is simple: every living creature is engaged in a full-time struggle to survive. The struggle operates on two levels—for the individual organism and for the species. The old stories of "fang and claw" and "the law of the jungle" seem exaggerated only because they are viewed from a human point of view. There is no high drama in the death of an animal when it is part of a natural process. Animals feed on animals, and for both the predator and the prey species this is salutary and important.

The predator is, so to speak, the stone on which nature sharpens the prey. In the long run a prey species not subject to natural predation would be in trouble. Weak, inferior and slow-witted animals would survive to breed and reproduce their kind.

As it is, only sure and alert animals have the opportunity to repeat themselves in ensuing generations in a natural system. Each animal that survives as a species must be considered successful in an eternal struggle. An animal that is not successful is dead and a species that is not up to the challenge is soon extinct.

Concealment and flight are means of escape. An impala, for example, need never chase anything, so its extraordinary speed and agility serve only a single purpose—avoidance of lion, leopard, cheetah, hunting dogs and hyena.

But survival of an animal as an individual is temporary. It is doubtful that any impala ever dies of old age. Sooner or later each individual must lose condition and slow down. Either it fails to detect a predator in time or fails to get clear once the foe is revealed. That doesn't matter, however, for the ability to survive is important only as long as the animal is of breeding age. The very old and the surplus young are no more than pawns in the scheme of things; experienced predators allow their young to practice stalking and killing them.

None of this was intended to please our sense of the esthetic. What is beautiful about it is the sureness with which species destined to survive function. A gorilla doesn't have to carry through an attack. Its show of rage is enough. The animal's rage, then, is a thing of beauty for its very magnitude does away with the need for violence. Similarly, young rhinoceros survive because the lion in Africa and the tiger in Asia, the only animals large enough to prey on even young rhinoceros, have had built into their own survival patterns the knowledge that a female rhinoceros is beyond their powers. A mother rhinoceros, therefore, need only display (we often call it bluffing) to drive away any cat that approaches her young.

In the surest sense, threat—bluff, if you will—and display are means of survival in the individual's and the species' battle to live. Animals that cannot mount a display that will intimidate or who cannot flee on winged feet must find alternatives. One of them is the masquerade of death. Almost everyone has heard that the opossum plays dead, but so do many other animals. The hog-nosed snake of North America, for example, will roll

Struggle for Survival

over and feign death if its huffing and puffing threat-display fails to halt any creature that threatens it. It has been suggested that animals that act out this charade of death may, in fact, go into a kind of shock and put on their act quite automatically. A number of birds feign a broken wing and act as decoys to draw predators away from their nests and young.

In a world that was soft and easy, without threat and competition, animals would lose their ability to adjust and adapt. Species would lose their tone. If there were no pressing environmental problems, mutations would remain random and offer no solutions. We can speculate that evolution would slow down and stop.

All creatures alive today are the products of all the pressures brought to bear on their ancestors since the beginning of life on earth. The pressures now being felt by the living creatures around us are forming the animals of the future. That is the continuum we observe in action.

Overleaf, preceding page: When threatened, a balloon fish (Diodon) *inflates itself by swallowing water. Removed from water, it does much the same thing, gulping air reflexively. Both air and water can be expelled at will. Its formidable array of spines lie flat against its body when it isn't inflated. Its defense mechanism is simple: it is painful to bite into and almost impossible to swallow.* [*Jane Burton: Bruce Coleman, Inc.*]

The moray eel (Gymnothorax), *above, is a vigorously territorial animal. Record lengths of ten feet have been claimed for them and their jaws are immensely powerful. They normally take up residence in a cave, reef or in bottom debris. Any intrusion is apparently taken to be hostile. Its defense is twofold: in stage one, shown here, it threatens to bite; in stage two it bites. Most animals are satisfied with stage one and so most morays go unmolested.* [*Ron Church*]

Display is a common technique when danger threatens. At left, in the Congo, a mantid waves its forelegs and arches its body back. Its size appears to increase and its posture denotes readiness to fight. In some species, warning displays are a bluff; in others they are backed up by real capabilities. [Edward S. Ross]
Above, a land-dwelling crayfish from eastern Australia (Enastacus spinifer) *exhibits its pinching claws. Two purposes are served as the animal elevates its body: it appears larger than it is and the claws are placed in a position to thrust inward and grasp on both sides. [Michael Morcombe]*

Above, a long-eared owl (Asio otus) *displays on a branch. The hunched-over position, puffed-out feathers, and elevated wings appear to increase the animal's bulk several times over. The glowering eyes in the facial disc add to the threat. The owl has powerful talons and a substantial beak; still, it is better to frighten an enemy away than risk a showdown. Adult owls have few enemies and most displays of this kind are territorial or mating in intent. They will also display over a kill.* [H. W. Silvester: Rapho Guillumette]

At right, an immature great skua (Catharacta skua) *rolls over and offers its claws to an intruder. Unable to escape, it threatens. Most predators large enough to handle a chick of this size are probably not put off for long. Its purpose may be delay. If the delay is long enough a parent may return and drive the intruder away.* [Eric Hosking: National Audubon Society]

The cottonmouth water moccasin (Agkistrodon piscivorus), *left, a relative of the rattlesnake, is highly venomous. Its venom, however, is primarily for food-getting and only secondarily for defense. It has been given another defense as a means of conserving venom: just as the cobra spreads its hood and the rattlesnake sounds its rattle, so the cottonmouth displays the white interior of its mouth. Most animals, including man, who live in cottonmouth country are familiar with the warning that that gaping white mouth presents.* [Jack Dermid]

Above, a frilled lizard (Chlamydosaurus kingi), *an Australian species not much over eight inches long, can erect a frilled collar seven inches wide. That proportionately enormous gape makes the lizard appear to be more formidable than it really is. Without the display it would be in trouble. Its bite, nonvenomous, would be of little consequence to anything large enough to be interested in eating it.* [Paul Popper, Ltd.]

Styles of passive defense: upper left, a Virginia opossum
(Didelphis marsupialis virginiana) *plays "possum." It is generally
said that such an animal is feigning death, but it is not clear why
this behavior would not merely make the animal more
vulnerable to some predators.* [Leonard Lee Rue III: National
Audubon Society] *The same question is raised by the seeming death
act of the eastern hog-nosed snake* (Heterodon), *lower left. The
snake first tried puffing and hissing and when that failed to
intimidate it flopped over into this death pose.* [Jack Dermid]
Upper center, a killdeer (Charadrius vociferus) *fakes a broken
wing and scampers away from its nest site in a noisy and obvious
manner. Its aim is to draw an attacker away and then take off.*
[Joe Van Wormer: National Audubon Society] *Lower center, a
nine-banded armadillo* (Dasypus novemcinctus) *rolls itself up into
a ball to protect its soft underbelly and present its hard-shell exterior
to an attacker.* [Leonard Lee Rue III: National Audubon Society]

Above, a gopher tortoise (Gopherus polyphemus) *also has a tough
exterior. Flight is impossible for so cumbersome a creature, but
the shell plus the ability to withdraw the head and neck and shield
them with the front legs make the creature a kind of living fortress.*
[John H. Gerard: National Audubon Society] *Overleaf: American
bison* (Bison bison) *bulls spar as the breeding season approaches.
This, too, is a survival device, for the stronger animal has
the best opportunity to breed. The bull at right will so
intimidate the lesser animal at left that it will withdraw and no
longer bother the stronger animal's mates. This kind of challenging
goes on from July to September. By eliminating weaker animals,
the species constantly strengthens itself with superior genetic
input. The young bulls stay with their mothers until they are
almost ready to breed. At first they are protected by the mature
bulls and cows, but as they approach breeding age they can
be killed by an enraged adult bull.* [Durward L. Allen]

This series shows the remarkable agility of desert rat kangaroos (Caloprymnus). Normally active only at night, these small rodents are preyed upon by a variety of reptiles, birds and other mammals. Their ability to survive depends on extremely sharp senses and instant reflexes. This fight (it may have been prompted by a female, food or just bad temper) demonstrates the fluidity of movement and the gymnastic potential of a species that must spend a good portion of its time avoiding being eaten. [Willis Peterson] The kangaroo genus Macropus (*pictured,* M. rufus, *the great red kangaroo), at right, contains the largest living marsupials. The males, robust, powerful animals, challenge each other and often fight although there is no set breeding season when this is most likely to occur. It is sometimes difficult to tell when a true challenge is being offered and met or when the animals are simply exercising or playing. Since great red kangaroos often live at least part of the time in fairly large groups, such fights are*

frequent. The agility of these animals makes a fight an active affair and this also may involve some serious pummelling because males may weigh almost 150 pounds. [*Harold J. Pollock: Alpha Photo Associates*]

Two other species in which the genetic potential is protected by male-to-male challenges. At left, two huge southern elephant seal (Mirounga leonina) *bulls fight it out on a beach. These animals may weigh as much as four tons and be over twenty feet long. They are the largest of all Pinnipedia, exceeding even the walrus in both weight and length. In this species the bulls assemble harems during the summer and fast rather than leave the females for even a few minutes. The bulls mature sexually at four, but it is usually one to three years after that before they are able to claim females and build a harem.* [Francisco Erize]

Above, two hippopotamus (Hippopotamus amphibius) *challenge each other in an African lake. Only the largest and most aggressive hippopotamus gain access to breeding females and fights often erupt among the bulls. Tourists return with photographs of hippos "yawning" when in fact they are challenging each other over territory adjacent to mature females. Males must prove*

themselves, must be the strongest of their kind. (It is worth noting how rarely females fight over males.) Evolution and the future of the species is protected in these sometimes bloody encounters between sexually aroused animals. [Klaus Paysan] Overleaf: Two immature red-tailed hawks (Buteo jamaicensis) *battle for possession of a carcass in the snow. In winter, survival can depend on being able to claim and defend food. Bluff works part of the time, but inevitably some challenges will go as far as actual combat. Birds such as vultures are used to sharing a carcass with others of their kind as well as other species, but not hawks. They will even kill to defend their claim. [John E. Swedberg]*

6

Once an object of any kind is illuminated, it is visible and recognizable in proportion to its dissimilarity to its background. It must stand out in some way for the eye to be able to separate it; color, texture, shape, some hint, some clue must be present or it vanishes into the overall visual experience that surrounds it. Sometimes the hint may be nothing more than the shadow cast by the object itself.

No matter what kind of eyesight an animal has, movement is probably the best single visual clue to another organism's existence. In the absence of movement the problem of visual detection increases enormously. Protective coloration, mimicry, cryptic design and behavior all have evolved along these principles. They tend to create a sameness, erase differences, minimize the effect of movement, blend the creature into its background and improve its chances for survival.

There are many ways in which protective coloration can work. It can combine with form and give us astounding examples of mimicry. Bugs can look like thorns, mantids like sticks, caterpillars like twigs, katydids like leaves, sweet edible bugs like violently poisonous or bitter ones. The possibilities have proved to be virtually endless.

Because photographers often tend to get the clearest views of their subjects and because, until recently, zoos offered few natural settings, the average animal viewer failed to appreciate the effect of animal markings. Unless you have been very close to a giraffe in the wild and failed to see it because its large splotches of color blended into the sun-dappled trees, you cannot understand its design. And what of the absurdly obvious stripes of the zebra? The zebra's arch enemy is the lion. You must always think of the lion when you try to understand a zebra's stripes. It was for the lion they were evolved, and to a lesser extent for the hyena and Cape hunting dog. The animals that prey on zebra are fairly low to the ground and when they look to the horizon, it isn't from our long-legged vantage point. The zebra's stripes provide visual discontinuity, create a disruptive pattern and obscure the animal's outline, its position and even the way it is facing. Confusion may not last long once the chase has begun, once a zebra herd begins to scatter, but in the world of the lion and its prey, the difference between life and death is measured in fractions of seconds. There is, too, the factor of a whole herd of zebra exploding into action. The visual confusion of all those stripes crossing and recrossing in front of each other must be bewildering for a stalker well below the level of the zebras' disruptively patterned backs. The leopard, the jaguar and the tiger are gaudy for another reason. They live among taller vegetation—trees, bushes, deep grass and reeds. Their coats must match the effect of filtered sunlight disruptively patterned by shadows. The great marked cats are not protectively colored so that they will not be eaten, but so that they will eat. This is as much a form of protection as that of a well-concealed prey.

Some animals are extremely dangerous to others but are better off when not forced to use their weapons. Even venomous snakes (who need their venom for food-getting) prefer not to be challenged. Many of them are well camouflaged. Camouflage, however, works just so far and no further. The generally dull cottonmouth water moccasin is well camouflaged to a point; when that point is passed the snake opens its mouth and throws its head back revealing the white interior. This advertisement tells all animals within range to beware. Some animals dangerous to others start out where the cottonmouth changes direction. They have warning coloration. Poisonous frogs are among the gaudiest and most conspicuous animals in the world. Their color says: "Don't eat me!"

And then there are the tricksters. Many moths and butterflies, and even some birds, have eyespots on their wings. They are

Camouflage, Mimicry and Other Built-in Defenses

diverters. Predators miss hitting vital areas on the first pass. Very often they don't get a second chance. Countershading is common among fishes. It makes a rounded surface look flat. Graded tones tend to eradicate a sense of depth. Surface and contour blend into the background.

It is generally agreed that an animal's eye is one of its most conspicuous features, and many tests have been devised to show that the round dark pupil of a snake or a fish tends to draw an observer's attention. No matter how well an animal is camouflaged, if the eye is conspicuous, the whole system is betrayed. For this reason many snakes, fish and even mammals and birds have eyestripes, dark markings that pass through the eye and break up its line. They spread it out, change its shape and reduce its magnetic effect. Eyestripes are found in hundreds of species.

Whether an animal has interruptive fringes to break up its outline, or advertises that it is dangerous (or pretends that it is by looking like something that really is), whether it is gaudy or subdued, its coloration, marking and shading are not gratuitous. Even when we don't understand how an individual pattern works, we can be certain that it evolved because members of the species that looked that way tended to live longer and breed more often. Almost nothing is gratuitous in nature. No aspect of animal existence would seem to bear this out more than color and form as they relate to survival.

Overleaf, preceding page: The sea slug (Chromodoris reticulata), *one of the nudibranchs or shell-less marine molluscs, is distasteful to fish, and they will not knowingly touch it. Its problem is to advertise its identity so that accidents will not occur. It does the nudibranch little good to be inedible if it is caught, crushed and then rejected. Its brilliant coloration is its warning. To taste awful and be immediately identifiable is the perfect combination in a world of eager predators.* [Jane Burton: Bruce Coleman, Inc.] *The pipefish imitates turtle grass, which it resembles in color, by assuming a vertical position and swaying with the water currents.* [Rober C. Hermes: National Audubon Society]

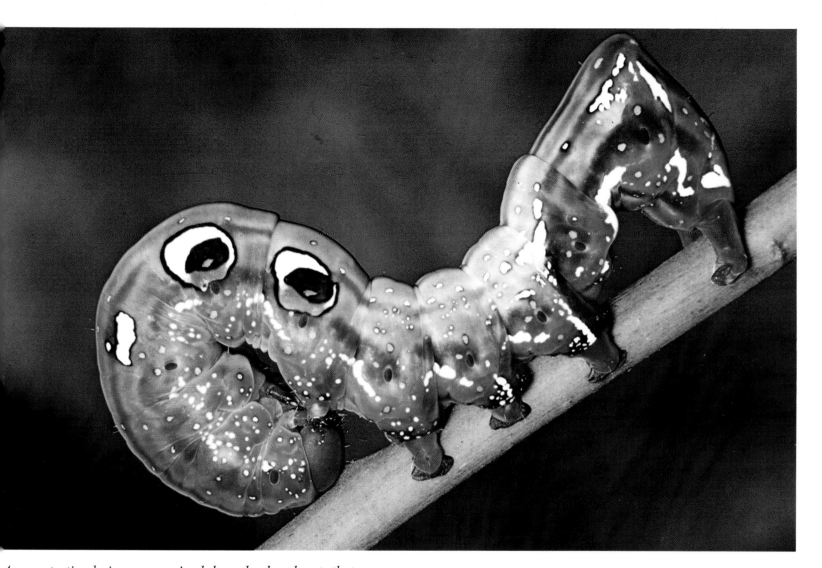

As a protective device, many animals have developed spots that
look like eyes. In adult butterflies and moths they are located
on the wings and probably serve to divert an attacker from
vital areas. On caterpillars, though, they are meant to intimidate.
Left, the larva of a spicebush or green-clouded swallowtail (Papilio
troilus) rests on a sassafras leaf. Huge eyespots give it a strong
resemblance to the smooth green snake (Opheodrys vernalis) and
may serve to frighten away would-be predators. The true
eyes are small and inconspicuous and are located on the underside of
the head. [Alexander B. Klots]
Above, the larva of a fruit-sucking moth (Othreis fullonia) has
eyespots on its side, again acting as a warning device, a bluff to
frighten enemies away by making the vulnerable animal seem to be
something it isn't. [Stanley Breeden]

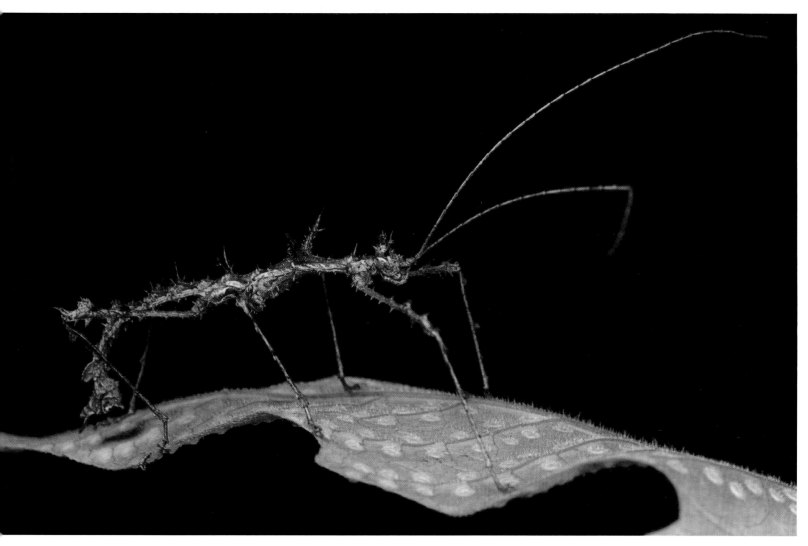

Spiny insects are protected by having a disrupted outline and indistinct shadows. The spines accomplish both these ends even when the insect is in full view. The spines may have a second protective function: giving pain to a predator. At left, from Ecuador, is a giant spiny katydid (Panacanthus cuspidatus) *and above is a spiny stick insect from New Guinea. Cryptic coloration is found throughout the animal kingdom, but cryptic form is far more common among the invertebrates. [Carl W. Rettenmeyer; Nicholas Smythe]*

The owlet moths belong to the large family Noctuidae, which contains over 20,000 species. At left is a Japanese form, Brahmaea japonica, *a night flyer so marked that it may well frighten predators away. In reduced light the eyespots could be intimidating. [S. C. Bisserot: Bruce Coleman, Inc.]*

Above is a Catocala concumbens, *pink underwing. These moths are the magicians of the insect world. Their top wings are cryptic and the animal simply cannot be detected against its background. It has, however, a brilliantly marked pair of underwings which it can expose at will, as the specimen here is doing. When pursued by a bird, the moth can land and show its underwings. As the bird approaches, the underwings are quickly covered and the moth seems to vanish. Alternatively, a moth that is being approached despite its cryptic wings can suddenly flash its brilliant underwings and confuse the would-be predator. The split second so gained can mean survival. [Alexander B. Klots]*

Animals that are repugnant to eat can afford to be conspicuous. The red arrow-poison frog (Dendrobatidae *spp.*), *above, exudes violently poisonous skin secretions. Animals attempting to eat it are not only repulsed, but may be killed.* [Alan Blank: Bruce Coleman, Inc.]

The grasshopper at right, Phymateus morbillosus, *is evil-tasting. Its brilliant red coloration serves to advertise its identity and keep accidents from happening. Edible animals are usually subtly colored—except for those that attempt to imitate these harsh-tasting species.* [Edward S. Ross]

Overleaf: *Sand-dwelling snakes are among the most successfully cryptic animals. Those that live in deserts are inevitably nocturnal and that adds to their capacity for concealment. Above, a dune viper* (Bitis peringueyi) *is caught by a flash camera at night in the Namib Desert of South West Africa. It is wiggling itself down into the sand where its color and texture will make it*

totally indistinguishable from its surroundings. Two purposes will be served: it will not be molested and therefore will not have to waste its precious venom defending itself, and it will be in a position to ambush small nocturnal desert rodents. [Edward S. Ross]

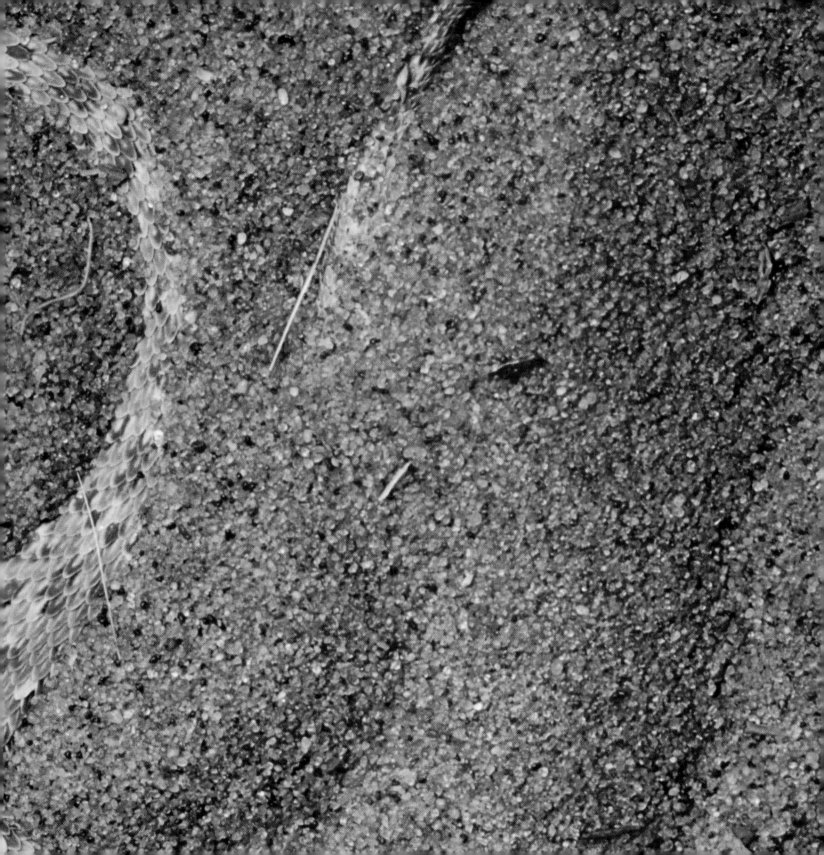

Some of the strangest colors and shapes are seen in the sea.
At left, a stonefish (Synanceja verrucosa) demonstrates how it blends
into the bottom rubble off an island beach. It is one of the most
venomous fish in the world but uses its toxin in defense only. The
venom glands are fixed to the fish's spines and only when the
fish is kicked or stepped on does it utilize its powerful toxin. It is not
clear why this species required both such cryptic coloration and
such toxic venom. The answer may be that it is a passive hunter: it
lies concealed in bottom debris and snaps up unwary fish that
happen too close.
Above, the bizarre Merlet's scorpion fish (Rhinopias aphanes)
sports a bewildering array of spines and protuberances. But, rather
than make the animal easy to see, this confusion obliterates it.
Where is its outline and where would one seize it? [Both by
René Catala, Aquarium de Nouméa]

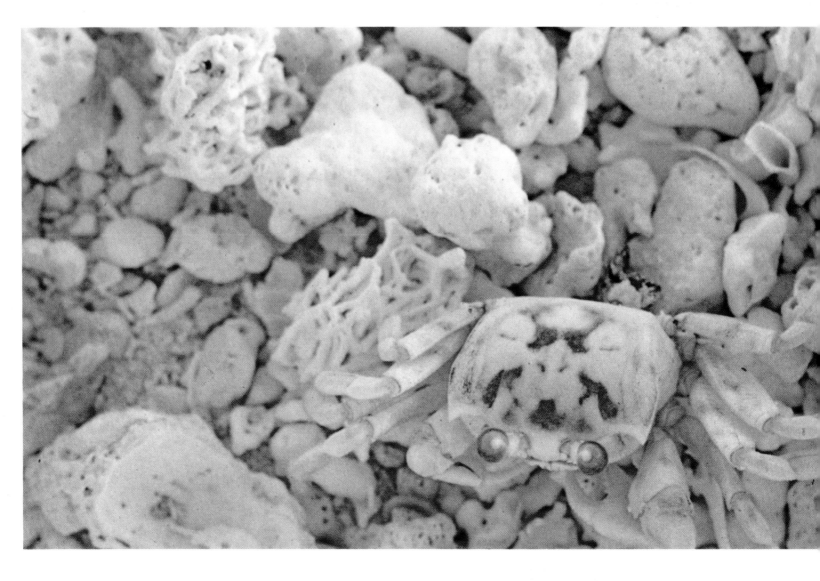

Amid coral debris and the skeletons of other living forms a ghost crab (Ocypodidae *spp.) picks its way. Potential prey to many animals, this little crustacean relies on cryptic coloration. It is only conspicuous when it moves, so it has learned to remain perfectly still, blending into its background. Even after a burst of activity that reveals its presence, it can vanish by coming to rest against confusing background materials. One cannot help being struck by the number of color combinations and markings that had to be tried before this one was achieved. Since evolution is continuous, this crab is probably only a step toward an even better color-marking combination. For the moment, though, we know things are working well, for it is surviving.* [David Cavagnaro]

Insects and other arthropods have an endless capacity for mimicry and other forms of concealment. Top left, treehopper nymphs of the family Membracidae mimic thorns so well that only the close-up lens or minute field observation can reveal them for what they are. [*Anthony Bannister: Natural History Photographic Agency*] Lower far left, the larva of a Papilio butterfly has been designed to look like bird droppings on a leaf. A resemblance to excrement is a common form of concealment among terrestrial invertebrates. [*Edward S. Ross*] Spiders have evolved such an appearance for two reasons: it fools enemies and attracts excrement-eating insect prey. Near left, in Panama, spiders of the family Thomicidae, the orchid spiders, achieve almost perfect concealment. [*Carl W. Rettenmeyer*] Above, a moth caterpillar stands out at an angle from the branch of a pine tree in Arizona and becomes, visually, a part of the plant itself. [*David Cavagnaro*] At right, a flower mantis (Hymenopus coronatus), *one of the many hunting mantids, positions itself at the end of a well-leafed stem and freezes, resembling the flower. The imitation may not be perfect but it can fool its prey long enough for the mantis to gain the advantage.* [*F. G. H. Allen*]

Other insect concealment techniques: at left is a devil's mantis (Idolium diabolicus) *looking like a half-eaten leaf. As long as the hunter stays still its prey will be unable to see it. When it finally moves it will be too late for any insect that has wandered near.* [Norman Myers: Bruce Coleman, Inc.]
Upper center, a pigmy locust in the Congo imitates a leaf sowell *that its concealment is nearly complete.* [Edward S. Ross] Lower center, a Hojarasca moth imitates a dead leaf. [Karl Weidmann]
Upper right, a Brazilian katydid, a female Cycloptera, *offers an astounding imitation of a local leaf, down to rust marks and spots of decay and insect damage. Just as amazing is the Peruvian katydid* (Augaura mirabilis) *at lower right.*
Among cryptically colored insects the survival of the species is *promoted by individuals that look most like their background. Given the rate at which insects reproduce, there are countless opportunities for trial and error.* [Both by Edward S. Ross]

At left, a white-tailed deer (Odocoileus virgianus) *hunkers
down in some vegetation and remains stockstill. A key factor
in its effort to hide itself is its ability to remain immobile for long
periods. Motion negates even the best protective coloration.
[Jack Dermid] Center, an alligator* (Alligator mississipiensis)
*appears to be nothing more interesting—or threatening—than a
floating log. It obviously hopes its potential prey will make such a
mistake. Above, a leopard* (Panthera pardus) *moves through
sun-dappled vegetation and is all but invisible. These spotted
and striped cats combine coat markings and stealth; they can move
at an agonizingly slow pace through concealing growth and not
give themselves away until it is too late—for their prey.
[Both by Leonard Lee Rue III: National Audubon Society]*

*An African lion can conceal itself by simply lying down in the grass.
Since a lion is built relatively close to the ground, disruptive
patterns are not as important as simple color blending. The grass
around it is honey-yellow much of the year, and so is the beast.
A lion has to work hard enough for a living without being
conspicuous.* [*Edward R. Degginger*]

Being changeably cryptic is a kind of peak in protective coloration. Here are three examples. At left is a ptarmigan (Lagopus). *Most species live above or beyond the tree line in a world where they are constantly exposed to danger. They have solved the problem by changing with the season—dark in summer, white in winter.* [Fred Bruemmer] *Center, a weasel* (Mustela) *displays the same capability. It probably serves a double purpose here, for the weasel is itself prey to birds of prey as well as being an active hunter of hares, rabbits and rodents.* [Karl H. Maslowski: National Audubon Society] *Above, a varying hare* (Lepus americanus) *changes its color with the seasons for the single purpose of not being eaten.* [Charlie Ott: National Audubon Society]

7

Although some plants can move—trees turn their leaves toward the sun, Venus's fly traps close on insects—and some animals such as sponges and corals are relatively stationary for much of their lives, motion is a characteristic of animal life.

Motion serves survival in many ways. Food-getting obviously requires the higher animals, at least, to move from place to place—to graze, to browse, to hunt prey. Motion enables male and female to unite and reproduce themselves in kind. Movement allows animals to seek optimum living conditions. A snake that could not move out of the sun and the cold would quickly perish. And movement on the grand scale we call migration enables some animals to adjust to the rhythm of the earth and follow the sun. Some animals move across incredible distances: the Arctic tern (*Sterna paradisaea*) may migrate over 20,000 miles every year of its life.

Man can perform some motions better than animals, but they are relatively few. We grasp with greater skill, and we manipulate the objects of our world with greater precision than do other animals. We walk upright rather more gracefully than the other primates (but not better, certainly, than an ostrich), but we fail in any comparison of speed. Ever since we rose up on our hind legs to free our hands and aid in our attack on other animals, we can be considered running animals—that is, we have been able to run as fast as is necessary to survive as a species. But when compared to the other running animals, how poorly we rate! Not all fast-moving animals can run, it should be noted. The elephant can move at 35 miles an hour, but it cannot run a step. It has only one gait, a walk, albeit a rapid one. It simply doesn't have the pulley/lever combination to gallop, trot, canter or jump.

Some animals are limited in the situations in which they can perform, while others are wonderfully versatile. A sloth is all but helpless in any position except upside down and in any place besides a tree. Most sea snakes are helpless on land. A gorilla has some reasonable agility aloft but it is hopeless in water. Tigers swim as well as some aquatic animals and no animal alive except a gibbon could catch a spider monkey in the trees. Cheetahs are the fastest four-legged creatures on earth (for short distances) in the open, but they could never maneuver among trees. Most fish are better off in water (but not all) and most mammals are better off out of it (but certainly not all!). Otters swim better than some fish (or they would starve) and bats (mammals) are more maneuverable in the air than many birds. There are, in fact, few firm rules about animal motion.

Most animals, of course, are of maximum interest to us when they are in motion, for it is when they are acting out their life style that we can observe their behavior. Fish in a stream or aquarium have a hypnotic effect because they are constantly in motion. We thrill at a horse on a track or at an impala gracefully arching away from a lion's plunging rush, and we are awed by the bluffing charge of a rhinoceros cow with her young nearby. A cheetah in a flat-out run is a marvelous sight, but so is a male butterfly performing over the head of a prospective mate or a gull coming into the wind. An albatross repeating the surface of the sea with its wingtip an inch from the water, a porpoise leaping clear, a pelican diving in, a shark fin slicing the water—all are wonders, all hold us spellbound. Animals move through the air, through the trees, across the ground, and through the sea. They burrow in earth, in wood and even stone. There is movement—visible and invisible—all around us. Our dance came from animal movement, and so did much of our earliest graphic art. An animal in repose may strike us as beautiful, as interesting, as fascinating even, but in motion many become awe-inspiring.

Movement

Overleaf, preceding page: A greater horseshoe bat (Rhinolophus ferrumequinum) *in flight—an active hunter that weighs less than an ounce. This small mammal normally hunts about twenty feet above the ground. When it catches a large insect on the wing, it lands to eat it. Its flight, not unlike a butterfly's, is fluttery. It can hover in midair. Its eyesight is poor, so flight after dark would be impossible were it not for echolocation: its huge ears enable it to maneuver past complex obstacles by catching the echoes from its high-pitched squeaks. Its squeals range from 50,000 to 80,000 cycles per second; the top range of the human ear is around 20,000 c/s. [S. C. Bisserot: Bruce Coleman, Inc.]*
In contrast to the swift, highly maneuverable flight of the bat is the agonizingly slow progress of an inchworm (Geometridae spp.) *toward the end of a twig. This inchworm will one day become a moth and have the power of flight, but in the larval stage it is limited to this looping gait because Geometrid moths have no larval prolegs (false legs used by the caterpillar to grasp leaves and hang on) on their third to sixth body segments. The lack of central legs requires this strange means of progression. [Gordon S. Smith]*

The seven-spot ladybird beetle (Coccinella septempunctata), *at left, engages in a burst of activity. Almost all of the 4000 species of these beetles are highly beneficial, feeding on many forms of insects harmful to plants important to man. Insects have about the same ratio of wing area to weight as birds do. The insect wing is a marvel of ultra-strong, ultra-light construction, far surpassing anything man can engineer.* [J. Markham: Bruce Coleman, Inc.]

A chameleon (Chameleonidae chameleo dilepis), *above, in an exposed position hurries to the security of a branch of a shrub or tree. Its feet, pattern of body colors and texture all render it at home in vegetation but not in the open. The feet of this small lizard are ideally suited for grasping branches and ill-adapted to movement across open ground.* [Edward S. Ross]

161

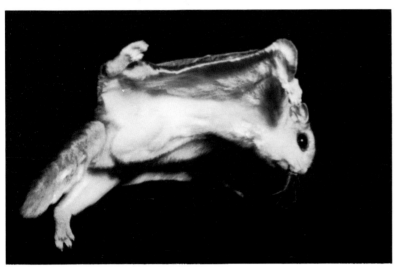

A flying squirrel (Glaucomys volans) *actually does not fly, but leaps from a high point, glides, and lands at a lower point. These squirrels have a furred gliding membrane extending along the sides of the body from the forepaws, or the neck, to the hind legs. The membrane, although very thin, is made up of sheets of muscles and can be relaxed and tensed as needed during the well-controlled glide. Just before landing, right, the little rodent lifts its tail, elevating the back part of its body. The landing is thus made straight on at reduced speed. Flying squirrels do not use wind currents the way birds do when they glide or soar, and therefore cannot normally gain altitude by gliding. [Leonard Lee Rue III]*

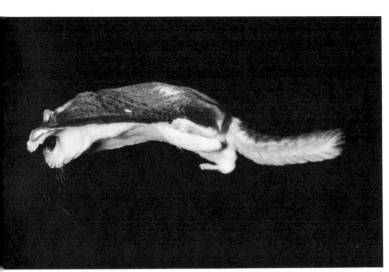

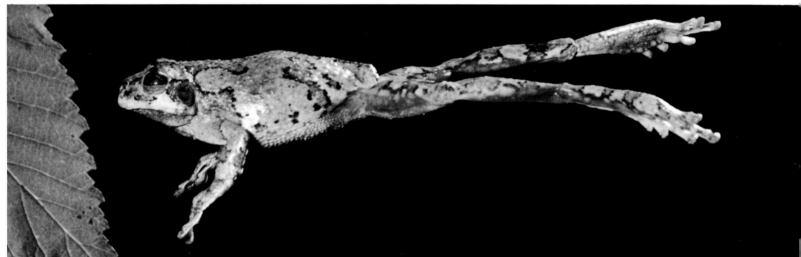

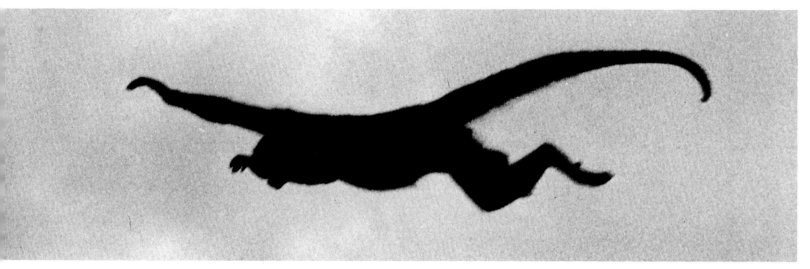

Three animals in a hurry, each equipped for speed under special circumstances. A spider monkey (Ateles geoffroyi), *at top, a prehensile-tailed primate of the New World, makes a spectacular leap from one tree to another. With the possible exception of the gibbon it is doubtful that any mammal is more agile aloft. The monkey's tail can be as long as three feet and acts as a rudder when the animal is in "flight."* [*Fran Allan: Animals Animals*]

At left, a gray treefrog (Hyla versicolor) *makes its acrobatic leap from branch to branch. As it becomes airborne it flattens its body and limbs into one plane to gain maximum gliding efficiency.* [*Jack Dermid*]

Above, a coyote (Canis latrans) *is seen in a flat-out run. This kind of speed, necessary for some hunting and for escape, is exhausting, and the animal is likely to become winded very quickly. Most wild animals do not have much stamina when pushed to such extreme exertion as that pictured here.* [*C. Allan Morgan*]

*A lion requires an enormous output of energy to launch
its four-hundred-pound body into a full run in a matter of seconds.
It can only sustain its full speed for a very short distance. [Leonard
Lee Rue III] The Thomson's gazelle* (Gazella thomsoni) *is
made for speed—to outrun lion and leopard and evade cheetah,
hunting dogs and hyena. A large part of the energy used by a
running animal on level ground is consumed in moving the limbs
quickly forward and backward: that is why gazelles and other
antelope have slender legs. [Alan Root]*

Three animal swimmers: the beaver, upper left, is a giant among rodents and may weigh as much as 65 pounds. It has webbed feet and its paddle-shaped tail, which may be sixteen inches long and five inches wide, serves as a rudder. It can remain submerged for fifteen minutes or more. [Jen & Des Bartlett: Bruce Coleman, Inc.]
At lower left is another superb swimmer, the hawksbill turtle (Chelona imbricata). *Despite its large size, it, too, is streamlined, with legs adapted into flippers for use in the sea. The front flippers serve to propel the turtle and the motion makes it look as though it were flying. [Norman Tomalin: Bruce Coleman, Inc.]*
Above, a Sergestid prawn cruises above a silty bottom 4000 feet down. The prawn can walk and swim both forward and backward. The forward motion is achieved with swimmerets, flattened appendages on the underside of the abdominal segments. Backward motion is accomplished with a convulsive flip forward of the tail fan. [Ron Church]

The flight of birds remains for man the most awe-inspiring form of animal motion. At left, a common tern (Sterna hirundo) performs its incredible hover, positioning itself while it searches the water for fish onto which it can plummet. When a bird hovers in quiet air it must place its body on a vertical plane and beat its wings backward and forward on a horizontal plane. Both strokes must be fully powered and this soon proves exhausting. [Gordon S. Smith] Upper center, a mute swan (Cygnus olor) literally runs across the surface of the water as it becomes airborne. Swans are among the heaviest flying birds, but are strong and graceful in the air. [Arthur Christiansen] Lower center, a Canada goose (Branta canadensis) lifts away with powerful downstrokes of its long wings. The enormous flight muscles to power these wings stretch halfway across the bird's chest to anchor on a keeled sternum. [Joe Van Wormer] Above, a long-eared owl (Asio otus wilsonianus) flies on silent wings. Three features work to keep the owl's flight quiet and facilitate its hunting. The upper surfaces of the wing feathers are padded with down so that they slide silently over each other. The feathers on the leading edge of the wing create a fine-toothed comb effect and, like the fringes on the trailing edge of the wing feathers, suppress all sound. [G. Ronald Austing]

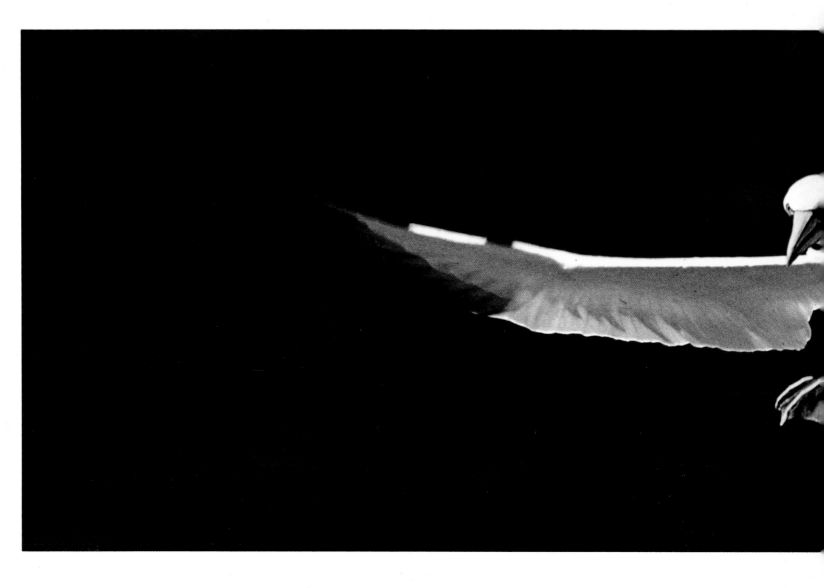

A gannet landing. Gannets are heavily built, with thick necks, short legs and large feet, but they are superb flyers and divers. They characteristically fly about 100 feet above the surface of the water and when they sight fish, drop into the sea in a plunging dive. When they brake for a landing they drop their long, wedge-shaped tails into a kind of scoop and begin to cup their long, pointed wings. They are not particularly graceful on land but in the air they are splendid. [*Les Line*]

Emperor penguins (Aptenodytes forsteri) *tobogganing. Penguins are the most highly adapted of all aquatic birds. Their wings are covered with small, scale-like feathers, giving the birds strong, narrow flippers. For increased speed, and perhaps just for a change of pace, they flop over onto their bellies and, using flippers and feet, slide along. They can pop back onto their feet again without apparent effort.* [*William R. Curtsinger: Rapho Guillumette*]

8

We have spoken of the individual animal as the means by which a species survives and goes forward in evolution. That can be too one-sided a description of life. An animal is not only a means to an end—the survival of the species—but an end in itself. Particularly in small groupings, family or otherwise, animals clearly interact on a one-to-one basis. Every lion in a pride has a place relative to every other lion in the pride. Similarly, a troop of baboons is an integrated and interrelated society. And the same is true of the other primates.

The nature of these relationships becomes more obscure when the group is so immense that each animal cannot possibly be aware of every other as an individual. Certainly, when tens of thousands, or even hundreds of thousands, of zebra and wildebeest mass for migration, the individual animals cannot be aware of any but those immediately around them. When animals mass for hibernation as snakes and insects sometimes do, the relationship each has to the other may not always be obvious. Yet those relationships exist. In masses of animals huddling or clustering against the cold there is a constant exchange of individuals between the cold and dangerous exterior of the mass and the warm, safe interior.

There is a critical mass for some species. Some species will not reproduce unless their numbers are astronomically high. The passenger pigeon, which became extinct in 1914, was not shot out of existence as is so often claimed. Millions were shotgunned, certainly, but millions more vanished because the flocks fell below their critical mass, the point below which the species could not reproduce itself.

Herds of animals that number in the hundreds of thousands are social units—or master social units, perhaps—made up of many smaller ones that have come together in a common pursuit. Harems, mated pairs, bachelor groups, unmated female groups and the many other groupings seen in various animals may blend together during migration. Surprisingly, even in the middle of incredible masses on the move, many of these relationships survive intact. A zebra on the edge of a band of migrating zebras a hundred thousand strong may not interact directly with a zebra on the other edge of the band, but it does with those closest to it, and they in combination with others, until the whole complex structure emerges. The word some scientists apply to massed herd behavior is *presocial*, meaning that it is something less than an organized social relationship.

There can be a certain amount of structure amid what appears to be total confusion. On East Africa's flamingo lakes millions of birds may feed together. At intervals along the shoreline can be seen dense groups of birds that seem more colorful than those around them. These tight little masses move as groups, heads going up and down in some mysterious rhythm. Always they are crowded in against each other. These are nurseries, young birds belonging to scores of parents but watched over by certain adults who hover around the edges. Amid millions of birds, "baby-sitting" details are formed. The same phenomenon is seen among penguins, which also assemble in enormous masses.

On the deck of a ship south of the Pribiloff Islands one can look in all directions and see nothing but the heads of fur seals bobbing in the water, tens of thousands of them heading for their own natal beach to give birth and breed again. There is organization here: every seal may not directly interact with every other, but unquestionably they are all affected by the mass.

Mass behavior helps species to survive in other ways. Large masses serve to protect individuals from predators that hunt along the migratory routes of most species. Weak and disabled specimens falling by the wayside are a kind of sacrifice for the safety of the rest of the herd. They draw predators away from healthy young, giving them a chance to mature. It is possible

The Individual and the Mass

that the massing of animals serves to spread disease and parasites, but if this were its only result, the practice would have died out. It is safe to surmise that some diseases among some animals, at least, are quelled by mass in ways that are still difficult to determine. And, above all, mass behavior does stimulate breeding and facilitate caring for young.

Large groups of animals, mammals, birds and often insects, are noisy places. There is an ambiance created within the mass as animals communicate by sound, call to each other, locate each other, warn each other, reassure each other. Communication is also visual and chemical. A mass of moving animals is an extraordinarily complex phenomenon and one that will provide us with subjects to study for years to come.

Overleaf, preceding page: The number of fish in a single school can be almost astronomical. Commercially they are reckoned in the tons. Here a school swirls through their twilight world as single organisms yet as parts of an organic whole with a life of its own. There is superb coordination: a threat to one side of the school will ripple through the mass, informing each animal almost simultaneously. They will react as individuals and as a school, responding as fluidly as the water itself. [Ron Church]

A small part of a herd of caribou (Rangifer tarandus) *on migration. Hundreds of thousands of these animals traditionally gather for their movement southward ahead of the Arctic winter. Although sexually segregated at certain seasons, caribou are highly gregarious. The individual needs of each animal and each group must be satisfied within the larger unit of the herd. In part, this life style has evolved in answer to the pressures of predation.* [Fred Bruemmer]

During migration the red-winged blackbird (Agelaius phoeniceus)
is extremely gregarious, thousands upon thousands gathering
for flight and feeding, as in this picture taken in New Mexico.
These birds range from Canada to Mexico. When each flock arrives
on its breeding ground, it breaks up as the males establish
territories and the nests begin to appear. After the breeding season
has passed, the flock again begins to assemble for feeding. The
birds moult and lose all of their tail feathers at the same time.
When new tail feathers grow back the flock is again ready to move.
What certain species gain from this extreme gregariousness is not
known. Related species like the orioles do not show this pattern.
[Thase Daniel]

Mexican freetail bats (Tadarida) *cluster on the ceiling of a cave. Millions of bats may use a single cave chamber and emerge in clouds at dusk to hunt the countryside for insects. They generally return to their cave before dawn and spend the day suspended in clusters like some strange fruit growing out of stone. Over the centuries the dung that accumulates on the cave floor can be many feet deep. Millions of bats moving in the restricted area of the cave as flight begins and ends are able to avoid collision through their echolocation capabilities.* [*Robert W. Mitchell*]

The large, colorful monarch (Danaus plexippus) *is the best known American butterfly. In the fall, monarchs congregate by the thousands for migration southward. They hibernate in a group, too, although the return trip in the spring is accomplished at a more leisurely pace and not in such vast numbers. Large groups of monarchs have been seen hundreds of miles at sea off both American coasts. It seems doubtful that they could have flown so far unaided, and yet we know that this species has migrated from Hawaii all the way to Taiwan in the west, and from the Eastern seaboard to the Canaries, the Azores and continental Europe.*
[*Jerry Gentry*]

Some animals survive only as members of a social unit in which they can claim a niche. If they can't function as a link in a chain, they fail as individuals. Some cannot live at all, or only briefly, without this association.

The most impressively society-oriented of all are the social insects—bees, wasps, hornets and termites. Some writers have described such colonies as organisms in their own right and likened the individual animals in a colony to corpuscles in a bloodstream. The specialization among social insects is extraordinary, individuals performing only certain of the functions of what we would consider a normal life cycle. Phases of their existence are allotted to others. In a beehive, for example, most of the inhabitants are workers—undeveloped females who can never reproduce. All the reproduction for thousands of females is done by a single developed female, called the queen, who does nothing else. Among ants and termites there are parallel systems. Some species of ants even develop special individuals called repletes, who act as living storage jars or honeypots for the rest of the colony to feed from in times of need. The repletes themselves are not eaten—just the nectar stored in their bodies.

There are no equivalents of the social insects among vertebrate animals, but even on the higher zoological levels we find deep-seated group instincts. We are not speaking of assemblages or super-herds of migrating strangers, as we were in the preceding chapter, but of animals interacting as "job-holders" within a community of animals.

Baboons assemble under a determined leadership and attack their arch enemy, the leopard. Lions cooperate in a hunt, one female driving hoofed prey into an ambush consisting of other females in the pride. Elephants are extremely herd-oriented. Females about to give birth are attended by other females; sometimes the young act as baby-sitters for even younger offspring; and a wounded animal is helped by herd-mates even when it means danger to themselves.

It is difficult to reconstruct the environmental forces that make a species social in nature, that select from mutations leaning in the direction of a tight society. What pressures did the African elephant experience, for example, not felt by the rhinoceros? The former is distinctly a herd animal, the latter is not. Why did the lion become the only cat to live in its particular kind of society? The tiger is closely akin to the lion yet it is solitary for most of its life. Not all bees and wasps are social; some are distinctly solitary. Why did some evolve into such social creatures and others in the opposite direction?

We cannot reconstruct the evolutionary choices made or discarded in each case but we can draw a basic inference. Becoming social was of survival advantage for the animals that took that path. It was equally efficacious, we assume, for others to remain solitary. Given all the pressures—competition within and among species, predation, food-getting, reproduction, suitable shelter—these two very different systems of life and relationships were in each case the right way.

As we said in the last chapter, we should distinguish between massed herd-migratory and truly social behavior. An assemblage of millions or perhaps even billions of locusts arrive at their flying stage at about the same time, and they feel the same pressures and respond the same way as they participate in massed swarming flight. Compare that with an anthill or a beehive built around a queen.

Or, compare Alaskan brown bears and wolves. Both are carnivores. Bears are essentially solitary animals for much of the year even though when blueberries ripen or salmon run, the great Alaskan brown bears will assemble along a salmon stream or in a blueberry patch. They manage there to sort things out after some squabbling and even an occasional death. But we cannot say

Cooperation

that there is a "bear society" except in the most temporary, loose sense of the term. Wolves, on the other hand, are extremely social. Normally, they do not fight among themselves. They most certainly have a hierarchal system, and there is some punishment, some demand for submission, and constant intimidation especially around mating time. True fighting, though, is rare. The alpha wolf is on top, the beta wolf is next in line, and their mates have corresponding positions within the patriarchy. In some situations, only the alpha wolf will mate. Ten wolves together in a hunting pack have a quite different relationship from ten bears in a salmon stream.

Cooperation and group behavior are ways of surviving. There are forces working against every species of animal on earth, forces that would wipe it out and make room for another. Those forces are what keep species on the move through evolutionary byways and into more beneficial patterns of structure and behavior. Group cooperation is one of the ways in which some species cope with those forces. Other species have found other ways. No computer will ever enable us to see the why and the how of those choices.

Overleaf, preceding page: Muskoxen (Ovibos moschatus) *form a defensive circle on the Arctic tundra. These feisty herd animals respond to threat with group action. The young are crowded into the center and the adults present to a marauder a wall of determination and horns. This defense worked against wolves, their traditional enemy, but men with rifles found it easy to kill an entire herd as they stood their ground. The muskox is now extinct in Siberia and Europe where it once ranged. There are remnants in Greenland and Canada and some transplants in Alaska. What may be a good survival idea in one era may contribute to an animal's decline in another. Nature did not anticipate killing for fun.* [Fred Bruemmer]

Canada geese (Branta canadensia) *generally use the V-shaped flying formation only on long flights. The lead goose is not only the guide but probably the strongest bird in the flock.* [Gordon S. Smith: National Audubon Society]

A herd of African elephants senses a possible threat and tries to locate it. The young have been crowded to the rear, trunks will come up, and the air will be searched for clues. Their poor eyesight will help them little. If they decide there is danger they will probably flee, marshalling their young ahead of them or on the side away from the threat. There may be an intimidating rush at the intruder that some would mistakenly call a charge. I have been "charged" half a dozen times; it was a bluff but it always worked. Elephants act in concert, which may help to explain their survival into modern times. [Leonard Lee Rue III]

Events in the life of the honeybee (Apis mellifera)*: A queen bee moves across a brood comb, left. Developing larvae can be seen in some of the cells. The queen is, as always, attended by workers. Upper center, the young workers hang together in chains on the combs or on the outside of the hive.*

Lower center, a foraging worker laden with nectar and pollen feeds a hungry worker that has remained on duty at the hive. By using their stingers, these defenders probably have in effect committed suicide. Generally, honeybees can use their stingers only once, leaving them in their victims. An exception may occur when the recipient is very soft and the stinger is not left behind.

Most ants are aggressive: here three small red ants have attacked a much larger wood ant. The outcome of such contests may be uncertain, but in this instance the attackers were victorious. Insects on the offensive differ from vertebrates in their inexorableness. You can frighten a lion from its kill and bluff a cobra into retreating; you can neither frighten nor bluff three small red ants. [*Four by Stephen Dalton: Natural History Photographic Agency*]

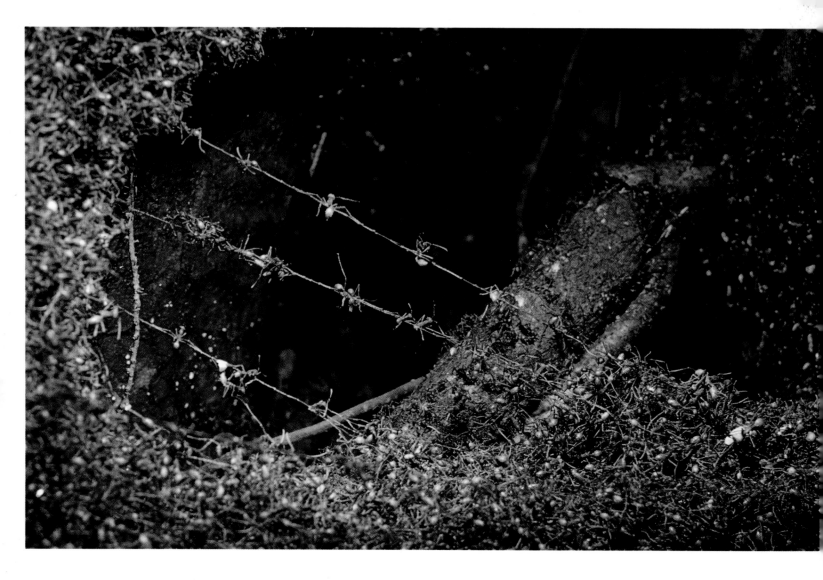

Army ants (Eciton burchelli) *at a bivouac during a period of migration. These nomadic ants form a living bridge across ant strands at the edge of the bivouac. An army of 150,000 or more of these animals may march each day, settling down in a relatively sheltered spot at night. The column during the day may be over 1000 feet long and anything that is overrun is eaten. Very few creatures can withstand the tearing action of the ants' jaws.* [*Carl W. Rettenmeyer*]

The only insects that rival, and in some ways surpass, the bees,
wasps and ants in social structure are the termites. But while the
other three groups are members of the order Hymenoptera,
the termites are in a much more primitive order, Isoptera, one they
have all to themselves. In the large picture above, a sand plain
in Australia is dotted with termite mounds or termitaria. In most
cases only the tips of the structures are above ground. Some
termitaria may go down fifty or sixty feet, weigh many tons, and
house millions of individuals.

At upper right, termites rush to close a rupture in the wall of their
nest. They are tireless, and repairing a cell is barely a challenge.
At lower right, a termite queen over four inches long is tended by
some of her offspring. Little more than a huge egg-laying machine,
she is a prisoner in her own empire, cared for endlessly by
hundreds of servants. [Three by Edward S. Ross]

Each species of animal has a unique life-situation. Natural selection has fostered the development of incredible adaptations to allow each species to make the most of the niche available to it. At times, two species evolve together; that is, a relationship develops between them that affects both of them. We speak of these rather special relationships as associations. Parasitism is by definition injurious to one member but beneficial, and often even essential, to the other. There is sometimes a fine line between animals that prey on others and those that parasitize. The difference is frequently determined by how lasting the association is.

Commensalism, which may be considered the opposite of parasitism, implies that one or both parties will gain from the association, and neither will suffer. When both gain, it is termed mutualism, and the animals are often said to be living symbiotically. By definition symbiotic does not signify a beneficial relationship, but the word is most often used that way. Commensal relationships usually involve mutual protection, or protection in return for shelter or food. Comfort might appear to be an advantage some species gain from an association. Certain fishes gain their food by cleaning parasites off other species; the species cleaned are made more comfortable, no doubt, but the animal's health, and hence its potential for survival, are what are really at stake.

The extremely venomous Portuguese man-of-war carries death for almost any small fish caught in its trailing tentacles. It has hundreds of thousands of stinging cells coiled and waiting to discharge when stimulated. But the small man-of-war fish, called *Nomeus*, literally hide out among the man-of-war's deadly weaponry. They certainly gain protection from the arrangement. Yet the Portuguese man-of-war rarely molests the *Nomeus*. Since these colonial hydroids have been observed eating *Nomeus* on occasion, it is apparently not distaste or danger to themselves that keeps them from feeding regularly on this small fish. The explanation, it is said, is that the *Nomeus* in fact does draw other fish into the range of the coelenterate's stinging tentacles. The communication between the *Nomeus* and the coelenterate is what is fascinating.

Sponges provide shelter to many small animals, and their porous structure accommodates and directs food-carrying currents to smaller animals of a number of species. One giant loggerhead sponge (*Speciospongia*) was found to be host to over 13,500 animals. Some of these animals, such as the shrimp *Spongicola*, the most abundant species found in the sponge, enter their host when small and are trapped there as they grow. That, truly, is a permanent association. It is difficult to see what the sponge gets in return except, perhaps, the substance of the shrimp when it dies and possibly some part of its excreta during life. Burrowing animals, both vertebrate and invertebrate, provide room and sometimes board to other animals. Prairie dogs find themselves sharing their burrows with burrowing owls, reptiles and a variety of insects and arachnids. An eagle may build a nest so large that it becomes a kind of apartment house: smaller birds nest in the lower layers and are not disturbed by the eagle. Some of them may help keep the insect and arachnid population of the eagle's nest under control.

The forms these associations take, their intensity and their importance to the respective species, all vary tremendously. There is no single way of viewing them or characterizing them. They are good for one or several species; they can be deadly to some. But somewhere some species must benefit.

Interaction

*Overleaf, preceding page: This large rhizostoma jellyfish or medusa
can kill the young jack* (Carangidae *sp.*) *that lingers nearby and
even brushes against it. The tentacles hanging below the bell of
the jellyfish contain tens of thousands of coiled stinging cells
which, when discharged, paralyze fish, allowing the jellyfish to draw
them in and digest them. Yet it tolerates the young jack and lives
with it in a kind of symbiotic relationship. The mystery is how
the fish identifies itself to the jellyfish. The jellyfish can't hear or see
it, yet it holds its fire. Does the jellyfish "taste" the fish or receive
chemical communications? If it does, how is the message passed
along to those stinging cells?* [René Catala, *Aquarium de Nouméa*]
*Many insects, particularly of the order Homoptera, secrete sweet
juices. Other insects such as the pastoral ants harvest these
juices. The pastoral ants stroke thorn-mimicking homopterans
with their feelers, and the mimics, in turn, yield a crop of
"honeydew" or nectar.* [Nicholas Smythe]

201

*A relationship in which one or both species benefit is called
commensal; when it is both we usually say the
animals are living together symbiotically. It isn't always easy to
be certain what the benefits are to both members of such a
relationship. At left, two pea crabs* (Pinnotheres maculatus), *living
inside the shell of a scallop* (Aequipecten irradians), *are
doubtless gaining a safe retreat, but what does the scallop gain?
Are the scraps of food torn up by the crabs utilized by the mollusc?
[R. N. Mariscal: Bruce Coleman, Inc.] Above, a sea anemone*
(Calliactus tricolor) *is attached to the carapace of a living
hepatus crab. Presumably the anemone's stinging cells offer some
protection to the crab while scraps from the crab's meals (crabs
are careless eaters) can be used by the anemone. Perhaps the
anemone "recognizes" by some primitive form of chemoreception,
taste or smell, an appropriate crab onto which to settle. [J. A. L.
Cooke]*

Mutualism can be a response on the part of two species to a negative situation one of them is experiencing. Here a wrasse (Pseudodax moluccanus), *right, is attended by a cleaner fish* (Labroides dimidiarus) *that is making its living by ridding the wrasse of troublesome but edible parasites—a common form of mutualism in the sea.* [Allan Power: Bruce Coleman, Inc.] *This relationship is also seen in land environments, as when oxpeckers, left, and tick birds, above, pick annoying parasites from hoofed animals on the African plains.* [Leonard Lee Rue III: National Audubon Society; Karl H. Maslowski]

A scene at an East African water hole: Burchell's zebra (Equus
burchelli) and white-bearded wildebeest (Connochaetes taurinus)
gather and drink. Typically, nearby, would be found Thomson's
and Grant's gazelles, and perhaps impala and wart hog. At
other times ostrich, elephant, rhinoceros and giraffe might mingle
with these animals. Such intermingling is random and accidental,
based on availability of water and food. Since water holes must
be visited regularly by most species they become favorite ambush
sites for predators. Often, herds of animals will stay back from
a water hole for hours before feeling safe enough to venture up to it.
When predators are hungry and hunting, every
animal in the area seems to know it. The way a herd of animals
moves up to a water hole is a good indication whether there are big
cats nearby. [Clem Haagner]

Many animals are able to survive in their natural environment by doing little more than eating whatever is readily available and by using existing shelters—a cave, a tree or a rock. Predatory animals require no auxiliary equipment. Their full hunting paraphernalia comes with the original animal itself. Browsing and grazing animals need only food and water within reach, their teeth, and occasionally their front feet for moving snow or loosening roots. Elephants have their trunks and some have tusks they can use to tip over entire trees.

But not all animals are so endowed. Some must manipulate substances other than their own body structure and the food they eat. We encounter among such animals some of the most complex behavior in the animal kingdom. It used to be said that man differed from all other animals in that he was a tool-user. This simply isn't so.

Chimpanzees select a twig and strip it of leaves and perhaps bark. They then thrust the twig into an ant nest, withdraw it when ants are crawling over it, and lick the insects from the tool.

They may repeat this process, using the tool over and over. True, they discard the tool when they are done ant-hunting, but proto-man and even early man tossed his rocks and clubs aside, too. So man is not unique as a tool-user; however, he is the *most advanced* tool-user. The idea that man alone can pass tool-using skills along to future generations may also not be accurate. Chimpanzees probably learn the ant-stick trick by watching other chimps do it.

There are different levels in manipulation, of course. The vulture that smashes an egg by throwing a rock at it engages in a far more complex behavioral pattern than a seagull that carries a clam aloft and drops it on a rocky shore. The latter is simply taking advantage of existing structures while the former is adopting a natural structure to its own special purpose.

Some animals create the substances they manipulate for survival.

Thus, many spiders produce their own silk for making escape routes, snares, nets, "parachutes" and other devices. Ants utilize silk produced by their larvae to sew leaves together. Trap-door spiders combine their own silk with debris to fabricate and camouflage their retreats.

Nests are created by a variety of animals. The vast majority of nest-building birds take advantage of local building materials—everything from mud to cast snake skins—but some (such as swallows) use saliva for cement and others (such as ducks) create linings for their nests out of their own downy feathers. The hummingbird steals spider silk and makes it into an incredibly soft, light nest.

Ants garden, growing fungi on beds they create out of other vegetable matter. Some keep slaves and some actually care for, transport and relocate insects that secrete sweet substances on which the ants feed. That is a form of manipulation, too—the manipulation of other organisms. Bower birds construct their display arenas out of local materials, and some painstakingly select objects for decoration. Some crush berries and smear the colored juice all over the bower. Were it not so obviously an anthropomorphic reaction, we could say that they were painting.

However much learning may be involved in animal behavior, the parallel with human behavior is accidental. If we try to understand a man manipulating things in his environment by watching a wasp build a mud shelter or chew wood for its nest, we shall go very far astray. For all the wonders we may behold in an animal's actions and reactions, no animal even approaches the flexibility, the open-endedness of human behavior. As far as we know, human capacity for learning and change is infinite. Animals are behaviorally finite creatures one and all.

Manipulation and Tool-Using

Overleaf, preceding page: An Egyptian vulture (Neophron percnopterus) *uses a stone to smash an ostrich egg. A number of birds use a rock or branch as an anvil against which to smash eggs, and seagulls carry clams up and drop them onto rocks and roads. But this is quite different: the vulture is using an outside object as an extension of itself. It has learned that a rock lifted in the beak and flung down is more powerful than the same amount of force directed through the unaided beak itself. It is using a tool.* [*Norman Myers: Bruce Coleman, Inc.*]

One of the most famous of the Galápagos Islands' finches that Darwin studied is the woodpecker finch (Camarhynchus pallidus), *above. It is a true tool-user. When pursuing insects that have retreated into a hole or crevice, it will use a twig as a probe to drive or bring the prey out into the open. The bird thus selects a tool of a weight it can manipulate, long enough to reach the desired depth, and not too wide. If, as has been reported by observers, the bird sometimes strips leaves from the twig before using it (in some instances, at least), then we may say the tool is not just selected and used, it is manufactured.* [*Tierbilder Okapia*]

One of the most extraordinary manipulations of the environment is performed by certain ants. Upper left, a brigade of Oecophylla workers draw the edges of leaves together. Other workers will appear with larvae in their mandibles and sew the edges together with silk produced by the larvae. [Edward S. Ross] At lower left, harvester ants sever the stamen from the flower of a hedgehog cactus. Some ants store plant food for later use (and constantly tend it) and others use it as a bed on which to grow fungi. The collection, transportation, storage and gardening of plant matter requires a great deal of labor and manipulation. Ants are masters at handling heavy and awkward loads. [William E. Ferguson] A tumble bug (Pinotus carolinus), above, digs its front legs into the ground and pushes a ball of dung with its hind legs. If it is early in the season the beetle will bury itself with the ball of dung and live off it as long as it lasts, then emerge and roll itself a new ball. Later in the season the ball will be buried, fashioned into a

chamber at one end and used to receive the beetle's egg. The larvae will live off the walls of its chamber, then pupate before emerging as a dung beetle itself. The balls, fashioned from freshly harvested animals droppings, are well rounded, so that the beetle can roll them a considerable distance. [*Edward S. Ross*]

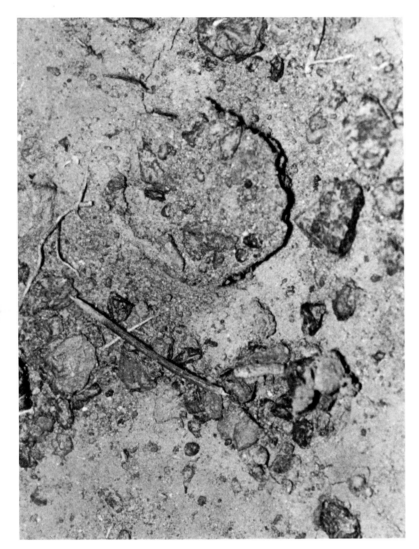

Left, the portal of the trap-door spider's secret den closed, and above, open. Spiders do more than spin webs. Some species excavate burrows, line them with silk, and glue a well-camouflaged lid together, attaching it to their den with an efficient silk hinge. In creating this small architectural marvel, spiders combine substances they produce within their own bodies (silk and adhesives) with sand and other ground debris. [Stanley Breeden]

At right, leaf-cutter ants carry their harvest back to their nest. Deep underground are chambers in which the cuttings will be stacked and where fungi will be grown to feed the colony. Here again, ants are the masters of manipulation. Between the time a leaf is selected for cutting and the fungi is harvested, many skills are involved. [Alan Root: Tierbilder Okapia]

Caterpillars, along with many other species of insects, can exude a silk-like substance that will enable them to cope with their environment. At far right, the larval instar of the hickory tussock

moth slips earthward on the silken strands it has created. The tensile strength of the silk is amazing, and the advantages to an animal that can produce its own escape routes are enormous. The upper end of the strand is glued to a tree limb or leaf with a substance also produced by the larva. [Gordon S. Smith] Overleaf: All spiders are predators. Many species actively pursue their prey, some jumping on them, others using an ambush. Many spin webs from silk they produce themselves. Web-spinners are virtually blind, despite their eight eyes, and rush to the point of disturbance only when the prey collides with their meticulously created snares. Here an orb weaver spider completes its stabilimentum or web-strengthening structure. Webs are placed where they are most likely to capture flying insects. [Edward S. Ross]

Epilogue

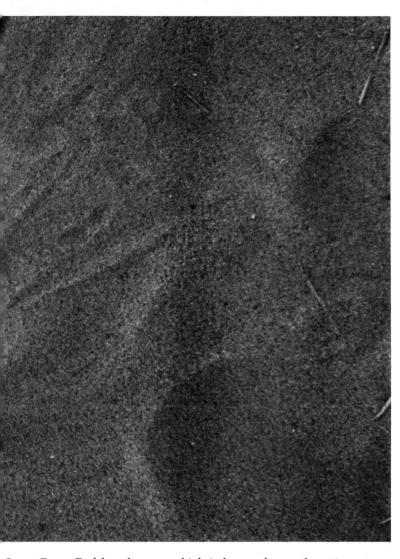

Man has **a** highly emotional, intensely personal view of death. He worries about the possibility of death and thinks about all deaths as if each were his own.

To an animal, death is less personal and never reflected upon. Birth happens, death happens; in terms of survival it is what takes place in between that matters.

The cosmos is a library of chemicals and all those borrowed must be returned. Man has tried to remove himself from this system by building mausoleums and encasing those who die in copper or bronze or steel. Animals die where they are, and through the metabolism of scavengers and the natural breakdown of other parts of their substance reenter the chemical cycle. There is little or no loss. The end of the individual cycle is a part of the larger cycle.

On a Cape Cod beach, over which it hovered countless times in the course of its life, lies this great black-backed gull, one of the largest and most aggressive of the Laridae. [*Gordon S. Smith*]
First overleaf: The end has come for an old cow elephant. The animals with whom she has lived mill around her, coax her, try to help her to her feet. Perhaps it has worked before. This time it will not. Strangely enough, when they pass this way again and find her bones, they will scatter them, carry pieces off into the brush and attempt to cover them with debris. [*Horst Munzig: Susan Griggs Agency*]
Second overleaf: The antlers of a dead caribou in the Far North. A time comes when an animal can no longer withstand the pressures of its environment—predators, parasites, disease, the extremes of weather. Returning to the earth, its body continues to serve in the endless cycle of life. [*Steven C. Wilson*]